Effective Learning in the Life Sciences

How Students Can Achieve Their Full Potential

Edited by David J. Adams

UK Centre for Bioscience, Higher Education Academy

WILEY-BLACKWELL

A John Wiley & Sons, Ltd., Publication

Library of Congress Cataloging-in-Publication Data
Effective learning in the life sciences : how students can achieve their full potential / [edited by] David Adams.
    p. cm.
    Summary: "Draws on experience from a major project conducted by the Centre for Bioscience, with a wide range of collaborators, designed to identify and implement creative teaching in bioscience laboratories and field settings"–Provided by publisher.
    Includes bibliographical references and index.
    ISBN 978-0-470-66156-7 (cloth) – ISBN 978-0-470-66157-4 (paper)
    1. Life sciences–Study and teaching (Higher) 2. Life sciences–Study and teaching (Higher)–Great Britain. 3. Creative teaching. 4. Biological laboratories. 5. Life sciences–Research.
6. Life sciences–Fieldwork. I. Adams, David J. (David James) II. UK Centre for Bioscience.
    QH315.E33 2011
    570.71'1–dc23

                                                                                                2011022847

A catalogue record for this book is available from the British Library.

This book is published in the following electronic format: ePDF: 9781119976653;
Wiley Online Library: 9781119976646; ePub: 9781119977636; Mobi: 9781119977643

Set in 9.5/11.5pt Times by Thomson Digital, Noida, India

First Impression   2011

To my colleagues in the UK Centre for Bioscience.
It was a pleasure and a privilege to work with you all.

Contents

List of contributors

David J. Adams
UK Centre for Bioscience, Higher Education Academy
Room 9.15, Worsley Building
University of Leeds
Leeds, LS2 9JT

Jo L. Badge
School of Biological Sciences
University of Leicester
University Road
Leicester, LE1 7RH

Lee J. Beniston
Leeds University Business School
Maurice Keyworth Building
University of Leeds
Leeds, LS2 9JT

Kevin Byron
The Learning Institute
Room 3.03A, Francis Bancroft Building
Mile End Campus
Queen Mary, University of London
London, E1 4NS

Maureen M. Dawson
c/o Centre for Learning and Teaching
Manchester Metropolitan University
2nd Floor, Cavendish North
Cavendish Street
Manchester, M15 6BG

Dawn Hawkins
Department of Life Sciences
Anglia Ruskin University
East Road
Cambridge, CB1 1PT

David I. Lewis
Faculty of Biological Sciences and Interdisciplinary Ethics Applied CETL
University of Leeds
Leeds, LS2 9JT

Martin Luck
School of Biosciences
University of Nottingham
Sutton Bonington Campus
Loughborough, LE12 5RD

Alice L. Mauchline
School of Agriculture, Policy and Development
University of Reading
PO Box 237
Reading, RG6 6AR

Stephen J. Maw
UK Centre for Bioscience, Higher Education Academy
Room 9.15, Worsley Building
University of Leeds
Leeds, LS2 9JT

Terry J. McAndrew
UK Centre for Bioscience, Higher Education Academy
Room 9.15, Worsley Building
University of Leeds
Leeds, LS2 9JT

Pauline E. Millican
c/o UK Centre for Bioscience, Higher Education Academy
Room 9.15, Worsley Building
University of Leeds
Leeds, LS2 9JT

Paul Orsmond
Faculty of Sciences
Staffordshire University
Mellor Building
College Road
Stoke-on-Trent, ST4 2DE

Tina L. Overton
UK Physical Sciences Centre
Higher Education Academy
Department of Chemistry
University of Hull
Hull, HU6 7RX

Julian R. Park
School of Agriculture, Policy and Development
University of Reading
PO Box 237
Reading, RG6 6AR

Julie Peacock
UK Centre for Bioscience, Higher Education Academy
Room 9.15, Worsley Building
University of Leeds
Leeds, LS2 9JT

Jon J. A. Scott
College of Medicine, Biological Sciences & Psychology
University of Leicester
University Road
Leicester, LE1 7RH

Joanna Verran
School of Health Care Science
Manchester Metropolitan University
Chester Street
Manchester, M1 5GD

Carol Wakeford
Faculty of Life Sciences
University of Manchester
1.124 Stopford Building
Oxford Road
Manchester, M13 9PT

Chris J. R. Willmott
Department of Biochemistry
University of Leicester
Leicester, LE1 9HN

David Adams was Director of the UK Centre for Bioscience, Higher Education Academy, from 2007–2011. Currently he is Director of Science and Research at Cogent Sector Skills Council.

Introduction

There has never been a more exciting time to study biology. We hear almost daily of major developments in new areas such as nanobiology, stem cell research or GM technology, and the popular media are forever running stories on the global impact of the biosciences. You have the chance to participate in this ongoing revolution, and if you are to make the most of this opportunity you must be prepared to think for yourself and fully engage in the learning process. If you can make this commitment then you should benefit greatly from this book.

Many of the book's contributors have interacted closely with the UK Centre for Bioscience, and its predecessors, during the last decade. Together they offer a wealth of experience and expertise in a wide range of areas of current importance in bioscience education. For the first time in a single volume, topics such as creativity, e-learning, bioethics and bioenterprise are considered, in detail, alongside more traditional elements of bioscience degree programmes such as laboratory classes and fieldwork. In addition, the book addresses areas and issues frequently identified by bioscience students as problematic. These include lack of confidence when using maths or stats in bioscience settings, difficulties when solving problems and frustration with assessment and feedback procedures. The book is designed to help you with these issues, and you will be able to access further support through an *Additional resources* section at the end of each chapter.

There is emphasis on interactivity, with inclusion of worked examples and case studies throughout. If you participate in these exercises and make the most of each chapter you will acquire a wide range of skills. These include many of the skills currently sought by prospective employers. Industrialists and university research laboratory supervisors alike indicate they want well-rounded graduates who can solve problems creatively in a wide range of settings. Enthusiastic engagement with the contents of this book should therefore help ensure not only that you benefit maximally from your time at university but also that you improve your employment prospects and achieve your true potential as a life scientist.

The book, chapter by chapter

Students are imaginative and inventive individuals, but unfortunately they are rarely given any help to achieve their true creative potential during bioscience degree programmes. A distinctive feature of this book is the inclusion of a chapter (Chapter 1) that will help promote your individual creativity and the creativity that often occurs when students work together in groups. As with creativity, students of the biosciences are given little help to develop their problem-solving abilities; in the second chapter you will therefore be shown how to approach algorithmic and open-ended problems with confidence. The next two chapters focus on practical skills in the biosciences with emphasis on students achieving their potential in laboratory and field. Continuing the practical theme, *in vivo* work (i.e. work with animals) is an area that has been identified as of paramount importance by the UK Government, researchers and educationalists, and an unusual and useful feature of this book (Chapter 5) is the consideration of a wide range of approaches and issues associated with the use of animals in the laboratory. In the final year of your studies you are likely to be engaged in a major research project. In recent years universities have offered a wide range of formats for projects, and these are considered in Chapter 6, which should help you identify the type of project best suited to your needs. The next chapter considers issues associated with maths and stats for biologists and describes how you can build your confidence in these areas. Chapter 8 contains a state-of-the-art update on e-learning in the biosciences, with advice on the use of new technologies including mobile phones, blogs, wikis, Facebook etc. You should know about traditional, as well as the most recent and innovative, assessment procedures used in universities. In addition you should be fully aware of the sort of regular feedback you can expect during your degree programme. These issues are considered in Chapter 10. It is essential that bioscientists should be able to communicate their ideas and general scientific information to other scientists and to members of the public. Chapter 11 describes traditional and novel approaches for communication in the biosciences. Two further notable features of this book are chapters on Bioethics and Bioenterprise. These are areas of great current importance to bioscientists. A considerable amount of material already exists in the field of bioethics, and Chapter 9 will raise your awareness of current approaches in this area. Bioenterprise and Knowledge Transfer are topics that are being embraced enthusiastically by many universities and the final chapter of the book considers how students of the biosciences can achieve their enterprising and entrepreneurial potential.

Tutor notes

The bioscience knowledge base is growing at a remarkable rate, and this can lead to tutors placing great demands on students who are asked to absorb enormous amounts of information. Unfortunately this can be at the expense of course components designed to promote independent thought and real engagement with the Scientific Method. This book is intended to redress this imbalance by raising students' awareness of their own considerable potential in areas of traditional and emerging importance. It is much more than a study skills guide, in that in each of the 12 diverse chapters the authors aim to build students' confidence to the point where they can decide for themselves whether they are making the most of their time at university.

You will find *Tutor notes* throughout, or at the end of, chapters. The notes will direct you to a great deal of additional material in support of teaching in the biosciences. This includes a very wide range of online and other resources provided by the UK Centre for Bioscience, Higher Education Academy.

David Adams
July 2011

1 Creativity

David J. Adams and Kevin Byron

1.1 Introduction

We should start by defining the terms 'creativity' and 'innovation'. *Creativity* involves original and imaginative thoughts that lead to novel and useful ideas. If you are to put these ideas to good use, you must be innovative as well as creative. *Innovation* may be defined as the exploitation of ideas in, for example, the development of new procedures or technologies. An excellent illustration of the distinction between creativity and innovation is the invention and development of the electric light bulb. Most people would identify Thomas Edison as the light bulb's inventor, yet over 20 individuals are thought to have invented similar devices up to 80 years before Edison's contributions. Only Edison was sufficiently innovative to refine his invention until it was a practical device that could be brought into commercial use in partnership with an electrical distribution company. You will learn how to ensure that *your* creative ideas are brought to fruition in Chapter 12.

Students of the biosciences are rarely encouraged to be truly creative or innovative (Adams *et al.*, 2009). A notable exception may arise during a final-year project, when you might be asked to come up with some novel ideas or solve a problem creatively. However, it is unlikely that you will be offered any help in generating original, imaginative thoughts or solutions. Indeed, in our view, bioscience students are rarely given the opportunity to develop anything like their full creative potential. This is a great shame because bioscience graduates will frequently be expected to be creative in a wide range of career settings.

In this chapter we consider a number of issues associated with the promotion of creativity in bioscientists. We start by inviting you to decide whether you consider yourself to be a 'creator' or whether your natural inclination is to be more of an 'adaptor'. The outcome of this exercise will help you make the most of the subsequent sections that deal with how to define problems, then solve them creatively as an individual or as a member of a team.

1.2 Adaptors and creators

It would seem that some people are naturally more inclined than others to take risks, challenge assumptions and be creative. Indeed, the psychologist Michael Kirton suggests that we can each be placed on a continuum based on our inclination to 'do things better' or to 'do things differently' and

Effective Learning in the Life Sciences: How Students Can Achieve Their Full Potential, First Edition.
Edited by David J. Adams.
© 2011 John Wiley & Sons, Ltd. Published 2011 by John Wiley & Sons, Ltd.

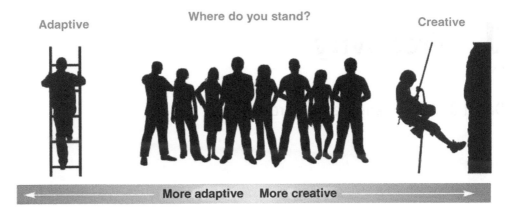

Figure 1.1   The adaptor–creator continuum (figure courtesy of Vitae, UK)

Table 1.1   Some of the characteristics associated with individuals located at the extremes of the adaptor-creator/innovator continuum

Adaptors	Creators/innovators
Seek solutions to problems in tried and tested ways	Willing to challenge the rules/assumptions and approach a task from an unusual angle
Risk averse	Like to take risks
Plagued by self doubt and always inclined to seek consensual view	Have low self doubt and pay little regard to the views of the majority
Focused, efficient, organised, disciplined with ability to concentrate on a task for a long period	Capable of detailed, routine work for only short periods
More likely to persist with a project and ensure outcomes delivered	May be poor at pursuing a project to fruition and ultimately making things happen

he labels the opposite ends of this continuum *adaptive* and *innovative*, respectively (Figure 1.1; Kirton, 1976). You may find it useful to consider where you fall on such a scale (Table 1.1). If you feel you are more of a creator/innovator than an adaptor then you are likely to benefit most from the problem-defining and problem-solving frameworks outlined in Section 1.3 of this chapter. On the other hand, adaptors should find of most value the techniques designed to promote creativity (Sections 1.4–1.9).

1.3   Defining problems

1.3.1   The 5Ws and 1H tool

Before committing a great deal of time and energy to creative problem solving you should make sure that you are entirely clear about the nature of the problem you wish to solve. The **5Ws and 1H** questioning tool may be used to help define and clarify the nature of the challenge.

Rudyard Kipling expressed this idea in verse:

I keep six honest serving-men
(They taught me all I knew);
*Their names are **What** and **Why** and **When***
*And **How** and **Where** and **Who**.*

Consider the example of a first-year bioscience student who is trying to decide whether she should apply for a placement in industry during degree studies. Perhaps most importantly, she begins by asking herself **Why?** she wants to do this, and concludes that she wants to acquire new skills, new perspectives on the science she is studying, contacts in industry and experience that will make her CV stand out in the crowd. Next she considers **What?** sort of work she would like to do and realises she would really like to work in a research laboratory. She wonders **Who?** will be affected by any decision to spend up to a year on a placement, perhaps hundreds or even thousands of miles from home, and realises that such a placement will have a major impact on friends and family. As a result, when she thinks about **Where?** she might spend the placement, she realises that this need not be in a large and distant company in the UK or abroad but could be in one of the much smaller companies located closer to home. She now thinks carefully about **When?** the placement should take place and compares the benefits of a formal, one-year placement with much shorter periods of summer or other vocational work. Finally she considers **How?** she can arrange for a placement ideally suited to her needs, and realises that she must build up a network of 'contacts' who can help her, including friends, family, her tutor and the University Careers Service. She also realises that she must 'target' companies engaged in the sort of research work she finds interesting and stimulating.

By weighing up all of the issues in this way, the student has defined, much more clearly, the problem she wants to solve. She now realises that she definitely wants the experience of working in industry during her three-year degree programme, but decides that she can obtain all the benefits she wants from interaction with industry by working during her summer vacations in one or more of the local 'spin off' companies associated with the universities located close to her home. The original problem: 'Should I apply for a placement in industry during degree studies?' has been redefined as 'How can I arrange for summer work in local biotechnology companies?' Her in-depth consideration of the issues means she is already well on her way towards solving this problem. However, now that she is clear about the real problem she wishes to solve, she may also benefit from engagement with the creativity techniques described in Sections 1.4–1.9.

1.3.2 Problem-solving frameworks

Various authors have devised fairly elaborate frameworks for creative problem solving, and perhaps the best known of these is the Osborn–Parnes creative problem solving (CPS) process illustrated in Figure 1.2. You will note that this framework has six steps involving objective, fact, problem, idea, solution and acceptance finding, and that each involves a period of 'divergent' thinking followed by a 'convergent' thinking phase. These terms should be defined at this stage.

If you are to be creative and have ideas, you must think *divergently* by using your imagination, challenging assumptions, rearranging information and examining it from new perspectives (see Section 1.4). Students of the biosciences are likely to be much more familiar with *convergent* thinking. It involves rational and logical reasoning that leads to convergence on the best solution to a problem. Convergent thinking is therefore essential when you wish to evaluate the ideas generated during a divergent thinking phase. In Section 1.3.1, the student who pondered how she might gain experience of industry was initially thinking divergently as she asked a series of questions,

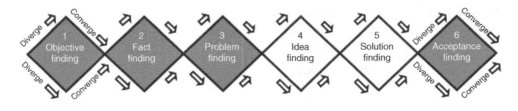

Steps	Stages
1–3	Identifying and clarifying the challenge
4–5	Generating and evaluating ideas
6	Implementing solutions

Figure 1.2   Osborn–Parnes creative problem solving

rearranged and re-examined information, and used her imagination. She then began to converge on the solution to her problem.

Creative problem solving is dependent upon an effective combination of divergent and convergent thinking: creative frameworks like the Osborn–Parnes CPS process (Figure 1.2) are designed to ensure that a period of divergent thought is always followed by convergent thinking as problems are defined, and ideas generated and evaluated. You can find out more about these structured approaches to problem solving elsewhere (see Section 1.14, Additional resources). We will return to convergent thinking towards the end of this chapter when we consider how you might evaluate your ideas. However, we will now focus on divergent thinking and the approaches you can use to generate the ideas you will need to solve problems creatively.

1.4   Accessing your creative potential

The approaches and techniques described in this section will help build the confidence you need to have ideas and solve problems creatively. If you are to be successful in this you will need to be bold and ready to move out of your 'comfort zone'. For example, you should be prepared to:

(1) **Welcome the unexpected:** Alexander Fleming noted a mould contaminant growing on his plated culture of bacteria. Instead of simply throwing away the plate, he looked more closely and observed inhibition of growth of the bacterium close to the fungal contaminant. He was curious about this effect and published his observation. During the next two decades Fleming's publication prompted others to isolate and develop penicillins as the first and, ultimately, most successful group of antibiotics. If, during research project studies, you notice something unusual, take the time to consider the implications of your observation.

(2) **Challenge assumptions:** during the 1968 Olympic Games, the American, Dick Fosbury, challenged the effectiveness of the popular 'straddle', 'scissors' or other high-jump techniques, and introduced the 'Fosbury flop' that involves the athlete jumping 'back first' over the bar. His willingness to challenge assumptions revolutionised the sport and helped win him a gold medal. A good and recent example of the importance of challenging assumptions in biology is provided by non-coding DNA. More than 98% of human genomic DNA does not encode proteins. Most of

these sequences have no obvious role and until recently were often referred to as 'junk' DNA. However, during the last few decades, many biologists have questioned the idea that such abundant, non-coding DNA should make no contribution to cellular activities in humans and other organisms. Their curiosity and investigations have been rewarded by the identification of an increasing number of diverse roles for non-coding sequences in gene expression, meiosis and chromosome structure, while additional lines of evidence indicate that other 'junk' sequences have essential but as yet unidentified roles in cells.

Unfortunately, students of the biosciences frequently require a great deal of encouragement before they will challenge assumptions. You should bear in mind that information provided by academics is not necessarily written on tablets of stone! This is of particular importance during lectures and seminars that involve cutting-edge developments in biology. In these situations you should keep an open mind and consider alternative interpretations and models that might be built around the data presented. Hold on to the curiosity about the natural world that probably led you to study biology in the first place, and don't be afraid to ask lots of questions!

(3) **Shift perspective:** when we shift perspective we change from one way of looking at things to another. In the illustrations in Figure 1.3 it is likely that initially you will be aware of only one interpretation. For Figure 1.3a you may see only a young woman wearing a feather boa, but if you look at the image in a slightly different way can you also see a much older woman? In Figure 1.3b you may see a pair of twins or a vase, but don't stop there. You might see a whale fin, a key hole, two cars parked bumper to bumper, a seal, a coat hanger etc. These are simple examples of what it feels like to shift from a single to an alternative, or multiple, perspective(s). If you develop the capacity to shift perspective then it is likely you will be more creative. The Hungarian biochemist Albert von Szent-Györgyi underlined the importance of shifting perspective in scientific research when he said 'Discovery consists of seeing what everybody has seen and thinking what nobody has thought'. He won a Nobel prize for his work on the isolation and characterisation of vitamin C, and another excellent quote attributable to this inventive scientist is 'A vitamin is a substance that makes you ill if you don't eat it'! Next time

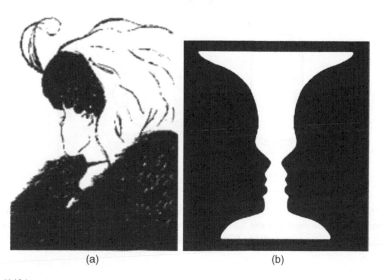

(a)   (b)

Figure 1.3 Shifting perspective

you have a problem to solve, try viewing the situation by looking at it from a different angle. For example, ask yourself how someone from another planet might solve the problem. Or, if you had unlimited money and resources, how that might make a difference to your approach. The new perspectives you adopt will hopefully help you be more creative.

(4) **Make connections:** look out for opportunities that will enable you to meet and talk with colleagues from other disciplines, e.g. chemistry, engineering. If a creative environment is designed carefully (see Section 1.7), it ought to facilitate this sort of interaction. You can then exchange ideas and perhaps identify unexpected connections between the problems you are trying to solve and what appear to be unrelated phenomena. An excellent illustration of the creativity that can emerge following the interaction of individuals with markedly differing backgrounds and expertise is provided by the Ultracane, a mobility aid for the visually impaired (Figure 1.4). It employs an echolocation technique similar to that used by bats (it therefore also provides a very nice illustration of 'bioinspiration' – see Section 1.5.3.4). The Ultracane came about through interdisciplinary brainstorming sessions involving Dean Waters, an expert on bats, Deborah Withington, a biomedical scientist with expertise in human aural physiology, and Brian Hoyle, an engineer and expert on intelligent sensing.

Figure 1.4   The Ultracane mobility aid: ultrasound transducers convert echoes from objects to vibrations in 'tactors' in contact with the fingers of the hand holding the cane. This, in turn, enables the brain to build a spatial map of the immediate surroundings (Figure courtesy of Professor Brian Hoyle, University of Leeds.*)

---

* www.soundforesighttechnology.com

Another good way to broaden your horizons and make connections is to attend obscure seminars that may appear, on the face of it, to be of only peripheral interest and relevance. You will be amazed by the new perspectives and insights these experiences can generate.

Fortunately, there are literally hundreds of techniques available that can promote creativity by encouraging individuals to challenge assumptions, shift perspective and, perhaps most importantly, make connections between what often appear to be unrelated phenomena. We describe a selection of these techniques in the following section and you will find many more in the books listed in Section 1.14, Additional resources, at the end of this chapter.

1.5 Creativity techniques

Well-managed, interactive group sessions (Section 1.8) can be extremely effective in generating ideas and suggesting novel approaches to problem solving. However, in a group, the views of the more dominant team members can rapidly prevail, and the potentially valuable thoughts and ideas of the more shy and reticent participants may be lost during discussions. In this section we therefore place emphasis on the generation of ideas by individuals *prior* to structured group sessions.

1.5.1 Case study Creativity in the Biosciences website

The freely available *Creativity in the Biosciences* website (Figure 1.5; www.fbs.leeds.ac.uk/creativity) uses a research-led teaching approach for the promotion of creativity in students working as individuals and in teams.

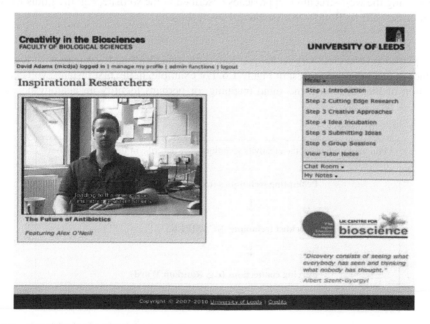

Figure 1.5 *Creativity in the Biosciences* website

- In short films experts describe cutting-edge developments in the biosciences, and problems associated with research.

- Working alone, students then access techniques that help them find creative solutions to these and other problems.

- They are encouraged to 'incubate' the problems and their ideas over a period of days.

- During this time they can use the website's Chat Room and 'Fridge Magnets' (electronic notice board) facilities to communicate their thoughts and ideas to other students in their team.

- Finally, the students come together in structured group sessions.

The approach adopted at this website ensures that extrovert students don't dominate proceedings from the outset. Instead each student is given the opportunity to be creative and contribute ideas that can be considered by all members of their team. The approach builds confidence and raises students' awareness of their individual creative potential.

1.5.2 Tutor notes

Individuals working alone often produce more and better ideas than they do when working in groups. Interestingly, the creativity of these solitary individuals can be further enhanced if they use computers to communicate and share ideas with other members of a group. Using a range of 'creativity techniques' at the *Creativity in the Biosciences* website, you can encourage students to explore their own creative potential and to communicate remotely with one another using chat room and electronic notice-board facilities. You can then facilitate interactive and creative group sessions using the well-structured approaches described at the website, e.g. the Lotus Blossom method or Edward de Bono's Six Thinking Hats technique.

We have placed several creativity techniques into three categories and you may wish to use these approaches in the order we suggest in Figure 1.6. For example, in the first instance you could try a combination of brainstorming and mind mapping, or perhaps identify an analogy between the

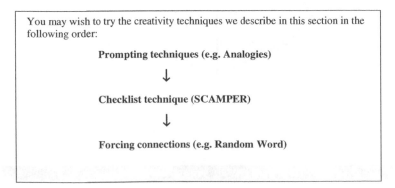

Figure 1.6 Using creativity techniques

problem you are trying to solve and a natural phenomenon, obscure activity or inanimate object. These techniques prompt 'associative thinking' in which you can make connections by following one word to another, associated word, phrase or image etc. and then build on any thoughts and ideas that may emerge. If you are stuck for ideas and can't think of any ways forward, then we suggest you try the SCAMPER 'checklist technique' which might help you come up with some ideas by asking a series of 'What if ...' questions. Finally, if all else fails, you should 'force' connections between the problem you are trying to solve and a random word or piece of information.

1.5.3 Prompting techniques

1.5.3.1 Individual/nominal brainstorming and purging

Carefully facilitated sessions of traditional brainstorming, which involve groups of individuals, can lead to the generation of many useful ideas (Section 1.8.1). In addition, you may find individual, or 'nominal' brainstorming a useful technique that will help you come up with new ideas and solve problems creatively. It's very simple: think of a problem you would like to solve, then, working on your own for a set time of (e.g.) 10 minutes, try to come up with as many ideas as possible. Anything goes, so write down every idea that occurs to you, no matter how apparently absurd or impractical – see the case study (Section 1.5.3.2).

Having purged your mind of what may be the more obvious ideas/solutions, you should now embark on the most difficult part of the exercise: allow another five minutes for further idea generation – you might try to come up with a further three to five ideas during this period. People often find that the most original and useful ideas appear during this final phase of a brainstorming session.

1.5.3.2 Case study   An outrageous idea leads to an original solution

There have been interesting instances of individuals coming up with ridiculous suggestions that led ultimately to very useful ideas. Paul Sloane (2009) relates the example of a company that packed delicate glassware and chinaware for dispatch. Newspapers provided cheap, convenient packaging materials, but the employees who packaged the goods stopped constantly to read the newspapers and the process therefore wasn't very efficient. Management convened a brainstorming session and a senior figure said 'Why don't we poke people's eyes out; they won't be able to see, and then they'll focus on packing.' This prompted someone else to suggest 'Why don't we employ blind people?' The company looked into this and it turned out that visually impaired people were excellent packers who were not distracted by the reading material. The company also benefited by improving its image as a socially responsible employer. All in all, an excellent illustration of how an outrageous idea can lead to a truly original solution to a problem.

1.5.3.3 Mind maps

Existing and emerging ideas can be captured in mind maps that will also allow you to identify novel connections during the creative process. Imagine you have been asked to devise a large university's strategy for effective communication of bioscience-related issues to the public. You should begin by

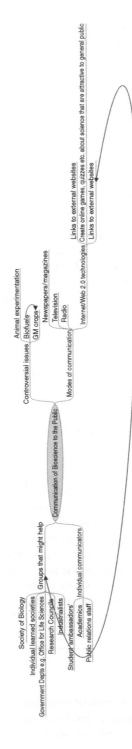

Figure 1.7 A mind map addressing a university's strategy for communication of bioscience-related issues

enclosing the problem in an oval in the centre of a piece of paper turned on its side. Next, brainstorm issues associated with the problem, for example: controversial issues you might need to address; individuals and groups who may be able to help; modes of communication etc. Draw lines out from the central oval for each of these topics (Figure 1.7). Mind mapping is sometimes described as 'visual brainstorming', and you can now brainstorm each of the themes identified in Figure 1.7 in greater detail and build further visual links to help you structure and develop your thoughts. For example, you might decide to seek help from various groups, and you record those societies and governmental organisations who should be able to make useful contributions. At the same time you use the mind map to record a range of modes of communication you wish to employ. This suggests further, online strategies (drawn as additional 'branches' on the mind map), and at this point you identify a link between the web pages on communication of science, which you will establish at your university, and the relevant websites of learned societies, research councils and government agencies. You then go on to consider individual colleagues who you might involve in this project. And so on.

All of this can be achieved by drawing a mind map on a piece of paper. If you adopt this approach you need not limit your mind map to a series of text boxes: people often find it useful to make sketches that represent ideas as they develop. Alternatively, there are several software packages that enable creation of mind maps on screen. These include the freely available FreeMind (http://freemind.sourceforge.net/); the mind map in Figure 1.7 was created using this software.

Mind maps can prove useful in a number of ways besides the recording and development of ideas. For example, many students use mind maps as aide-memoires, and you may find them invaluable as revision tools when preparing for exams.

1.5.3.4 Analogies

In an analogy, two things, which are essentially different but which nonetheless have some similarities, are compared. If you can identify an analogy between a problem you are trying to solve and a distant phenomenon or activity, you may find that the analogy suggests a creative solution to the problem. When NASA needed to retrieve a satellite tethered to a space station by a wire 60 miles long, they found they could not simply reel-in the satellite; if they did so, the satellite would begin to act like a pendulum, dangerously swinging with an increasing arc. They came up with the analogy of a yo-yo and applied it to the problem. The solution involved fitting a small motor onto the satellite itself. The satellite used the motor to crawl back to the space station – like a yo-yo along a string.

A number of very important inventions and breakthroughs have occurred as a result of an analogy drawn between a problem and a natural phenomenon. An early example of this 'bioinspiration' was the invention of the telephone, in the nineteenth century, by Alexander Graham Bell. Key components of Bell's transmitting and receiving devices were artificial, vibrating diaphragms that converted changes in sound to changes in electric current or vice versa. They were developed following his identification of an analogy between these structures and the vibrating, tympanic membrane of the human eardrum.

A more recent application of bioinspiration in creative problem solving was the use of 'platelet' technology to repair damaged oil and water pipes. In the human body, platelets aggregate at sites of injury and have a key role in plugging gaps in blood vessels, thus making a major contribution to tissue repair. In 1998, Dr Iain McEwan, an academic at the University of Aberdeen, proposed that a similar approach might be used in the repair of damaged oil or gas pipelines, which would involve little or no disruption of production. The 'platelets' his spin-out company, Brinker Technology, developed are free-floating, discrete particles that, when injected into a pipeline, are transported downstream with the flow until they reach a leak where the escaping fluid forces them into the gap, thus providing a seal. In another interesting development, the UK utility company Yorkshire Water is currently evaluating the effectiveness of platelet technology for the repair of leaking water pipes.

Perhaps the 'platelets' approach could be extended based on the identification of a further analogy between the capacity of human platelets to attract and interact with other cells, and maintenance of a pipeline that has been repaired using 'platelet' technology. The 'platelets' used for the initial repair of the pipeline could be coated with a substance that has a strong affinity for a second batch of 'platelets' released into the oil or water pipeline. Interaction of the first and second batches of platelets ought to reinforce and consolidate the initial repair. You will find further examples of bioinspiration in the books listed in the 'Additional resources' section at the end of this chapter.

Look out for a source of bioinspiration by identifying an analogy between the problem you are trying to solve and a natural phenomenon. Alternatively, you may find inspiration from an analogy that involves an unusual object or an obscure activity.

1.5.3.5 Visualisation

We have already seen how the same image can be interpreted in a number of ways (Section 1.4), and that the identification of multiple perspectives can help promote creativity. Visualisation of images can, in its own right, be a powerful tool for idea generation. A very nice illustration of this was provided by Michalko (2006), who relates how Karl Kreckman, a chemist working on the protection of corn seeds from harsh environmental conditions, was inspired by his son's drawings. One was a picture of a tree wearing a fur coat and hat. This made him think of seeds protected by clothing and the nature of the synthetic materials used to make clothing. These include polymers, and this thought process led to his idea for intelligent polymer seed coatings that allow seeds to be planted in any weather or season. The seeds remain dormant under cold, non-optimal conditions, but the properties of the polymer coating change as conditions improve, permitting the seeds to germinate when the soil reaches the right temperature.

Try summarising a problem you wish to solve by sketching one or more images. Many people are reluctant to do this because they feel they lack the artistic talent, but this doesn't matter. Creating even the simplest of images should stimulate the imagination. Furthermore, the level of 'relaxed attention' associated with drawing or painting will help achieve a degree of detachment conducive to creative problem solving.

1.5.4 Checklist technique

1.5.4.1 SCAMPER

SCAMPER is a particularly effective creative problem solving technique that helps generate ideas through the identification of new perspectives. It is an acronym for a checklist of 'What if we Substitute, Combine etc.' questions as follows:

Substitute	What else could we do/use instead; what aspects of a procedure, components of a product, elements of a role might be replaced?
Combine	Blend methods, materials, ideas that are usually separate.
Adapt/**A**djust	Alter the nature of an existing procedure or product, or copy it but in the new context.
Modify	Change meaning, size, frequency etc.
Put to other use	Re-purpose procedure, materials etc.

Eliminate   Divide, separate, subtract, delete a major element or function.

Rearrange or Reverse Do the opposite, work backwards, arrange in a different order.

Here is an example of how the SCAMPER technique might be used in the development of a university's policy for communication of science to the general public (see also Section 1.5.3.3):

Substitute   Recruit students, rather than experienced, professional scientists, as 'ambassadors' for communication of key issues to the public.

Combine   Take the opportunity to communicate exciting developments in science to school students during school recruitment visits.

Adapt/Adjust   Instead of focusing on interaction with older children and adults, develop a programme that communicates key issues to children of primary-school age.

Modify   Don't rely on static displays during exhibitions or conferences; instead use interactive, 'hands on' materials.

Put to other use   Use the materials developed for communication of science to showcase the activities of science departments during visits to the University by potential funders of research, alumni, prospective students, their parents etc.

Eliminate   Keep it simple: ensure that members of the public are not overwhelmed by information. Focus on a few key areas and ensure that the underlying science and its significance are communicated clearly.

Rearrange/Reverse Ask members of your audience to describe their perception/ understanding of key issues, then engineer a discussion/debate based on the points that arise.

Try applying the SCAMPER 'What if . . .' questions to a problem of your choice. It's very unlikely that answers to each of the seven SCAMPER questions will immediately lead to a novel idea. However, if one or two of these prompts help identify new insights and suggest creative approaches to problem solving then the technique will rapidly prove its worth.

1.5.4.2 Case study SCAMPER and New Scientist

A single issue of *New Scientist* (26th September, 2009) described a wide range of breakthroughs/ developments/innovations that could have resulted from the application of the SCAMPER technique.

Substitute: *'Virtual cop to run identity parades'* – computers substitute for detectives in guiding witnesses through identification of suspects.

Combine: *'Drug-electricity combo makes paralysed rats sprint'* – a combination of drugs and electrical impulses allowed rats with severed spinal cords to sprint.

Adapt: *'Locust flight simulator helps robot insects evolve'* – adapt insect flight aerodynamics for the creation of the ultimate surveillance machine.

<u>M</u>odify: *'Entire cities recreated from Flickr photos'* – <u>m</u>odified to create a virtual model of Rome.

<u>P</u>ut to other use: *'How nuclear power could fertilise fields'* – <u>p</u>ut waste heat generated by nuclear reactors to (green) use in manufacture of fertiliser and generation of hydrogen.

<u>E</u>liminate: *'Mutant mice live the dieter's dream'* – <u>e</u>limination of obesity gene regulator caused mice to eat less but retain normal metabolic rate.

<u>R</u>everse: *'Bashed your head? You needed a stiff drink'* – in <u>r</u>eversal of conventional wisdom, consumption of alcohol may improve outcome for brain injuries and ease sore heads!

Whether or not the individuals who came up with these ideas utilised the SCAMPER procedure to prompt their creativity is unclear. Indeed many creative leaps occur subconsciously, but there is no doubt that the SCAMPER tool can be applied successfully in a conscious manner if inspiration for ideas is not forthcoming.

1.5.4.3   Tutor notes

A number of authors describe SCAMPER in much greater detail. For example, Michalko (2006) devotes more than 40 pages to this technique.

You may wish to consider one or more of the 'What if . . .' questions in greater detail. For example, for **Combine**, it's worth stressing that combinations of disparate methods, materials, ideas etc. often led to major developments and discoveries. An excellent example is the invention of the printing press, during the fifteenth century, by Johannes Gutenberg, who combined the power of a wine press with the flexibility of a coin punch. A more modern example is the clockwork radio, which is a combination of a clockwork dynamo and a radio.

1.5.5   Forcing connections

1.5.5.1   Random word

If you are making little progress with the other creative problem solving techniques described in this chapter, try to force connections between the problem you are attempting to solve and an item chosen at random. The novel associations that result may identify fresh perspectives so that you approach the problem from a new direction.

A very simple way to do this is to choose a word, at random, from a list. Try this with the words listed in Figure 1.8 by closing your eyes and pointing at the list with a pen. What does the word you have selected make you think of? Can you identify any association with the word, the train of thought it evokes and the problem you are trying to solve? Most of the words in the list will probably not lead to the generation of useful ideas, but some may. For example, imagine you have been elected a student representative for your faculty's Learning and Teaching Committee. The committee discusses how the faculty might do more to help students realise their true potential (a major underlying theme for this book), and members of the committee are asked to bring their

Adult	Circle	Gemstone	Passport	Spiral
Aeroplane	Circus	Girl	Pebble	Spoon
Air	Clock	Gloves	Pendulum	Sports car
Aircraft carrier	Clown	God	Pepper	Spotlight
Air force	Coffee	Grapes	Perfume	Square
Airport	Coffee shop	Guitar	Pillow	Staircase
Album	Comet	Hammer	Plane	Star
Alphabet	Compact disc	Hat	Planet	Stomach
Apple	Compass	Hieroglyph	Pocket	Sun
Arm	Computer	Highway	Post office	Sunglasses
Army	Crystal	Horoscope	Potato	Surveyor
Baby	Cup	Horse	Printer	Swimming Pool
Backpack	Cycle	Hose	Prison	Sword
Balloon	Database	Ice	Pyramid	Table
Banana	Desk	Ice cream	Radar	Tapestry
Bank	Diamond	Insect	Rainbow	Teeth
Barbecue	Dress	Jet fighter	Record	Telescope
Bathroom	Drill	Junk	Restaurant	Television
Bathtub	Drink	Kaleidoscope	Rifle	Tennis racquet
Bed	Drum	Kitchen	Ring	Thermometer
Bee	Dung	Knife	Robot	Tiger
Bible	Ears	Leather jacket	Rock	Toilet
Bird	Earth	Leg	Rocket	Tongue
Bomb	Egg	Library	Roof	Torch
Book	Electricity	Liquid	Room	Torpedo
Boss	Elephant	Magnet	Rope	Train
Bottle	Eraser	Man	Saddle	Treadmill
Bowl	Explosive	Map	Salt	Triangle
Box	Eyes	Maze	Sandpaper	Tunnel
Boy	Family	Meat	Sandwich	Typewriter
Brain	Fan	Meteor	Satellite	Umbrella
Bridge	Feather	Microscope	School	Vacuum
Butterfly	Festival	Milk	Sex	Vampire
Button	Film	Milkshake	Ship	Videotape
Cappuccino	Fingerprint	Mist	Shoes	Vulture
Car	Fire	Money $$$$	Shop	Water
Car race	Floodlight	Monster	Shower	Weapon
Carpet	Flower	Mosquito	Signature	Web
Carrot	Foot	Mouth	Skeleton	Wheelchair
Cave	Fork	Nail	Slave	Window
Chair	Freeway	Navy	Snail	Woman
Chessboard	Fruit	Necklace	Software	Worm
Chief	Fungus	Needle	Solid	X-ray
Child	Game	Onion	Space shuttle	
Chisel	Garden	Paintbrush	Spectrum	
Chocolates	Gas	Pants	Sphere	
Church	Gate	Parachute	Spice	

Figure 1.8 Random word list

ideas to the next meeting. You decide to use the 'random word' approach and select the word 'fingerprint', at random, from the list (Figure 1.8). Fingerprints are unique to each individual, and this makes you think about how each individual student also has a unique set of talents. Not all of these talents will necessarily be readily assessed using traditional approaches like essay writing or laboratory write-ups. You therefore decide you will suggest that the faculty should introduce a wider range of assessment procedures that might involve, for example: completion of open-ended problems or tasks that may have more than one solution or outcome (rather than typical assignments that usually have only one correct answer); students designing experiments (rather than simply following standard procedures outlined in laboratory manuals); or completion of reflective portfolios, over e.g. a semester, in which students are expected to demonstrate how the learning objectives for modules have been met. These alternative procedures will give students additional opportunities to do well, build confidence and become more aware of their true potential.

1.5.5.2   Ideas in a box (also known as Matrix analysis or Morphological analysis)

This technique was developed by Fred Zwicky (1969). It prompts ideas by forcing together the *attributes* and *sub-attributes* of a problem. For example, when a group of students was asked to develop a novel approach for the diagnosis of disease, they began by listing attributes and sub-attributes of this problem, as shown in Table 1.2. The solution they came up with was to create live 'biosensors' incorporating genetically modified bacteria. These organisms would be designed to respond to the presence of very small amounts of disease-associated molecules by secreting much larger amounts of indicator molecules that could be detected in a colorimetric reaction, or using a biosensor on an electronic 'chip'.

Their idea stemmed from a combination of the sub-attributes 'living' (**Attribute: 'Materials made from'**) and 'biosensors' and 'colorimetric assay' (**Attribute: 'Techniques that might be exploited'**) (Table 1.2). They felt their approach could be used for the diagnosis of a very wide range of diseases not only in humans and animals but also in plants and other organisms. In addition, some of the other attributes/sub-attributes might suggest, for example, that live biosensors could have applications in drug testing of athletes, forensic medicine or the detection of toxins released by bioterrorists. The sub-attributes for the attributes **Communicating information** and **Added functionality** suggested many options for additional features of any new device developed. And so on. Of course, the idea of live biosensors may prove to be entirely impractical; the example has been outlined here merely for illustrative purposes.

You should try out the technique for yourself. Identify a problem you are keen to solve and spend some time constructing a grid. Be sure to include some unusual features for each set of sub-attributes. As we have indicated elsewhere, it's often the more outlandish suggestions that lead ultimately to interesting and hopefully useful ideas.

1.6   Incubation

The deliberate incubation of a problem can take a little practice. Most people would, naturally, prefer to solve their problem instantly, but when you are looking for creative solutions it pays to withdraw for a while and trust in an incubation period. This strategy has proved successful for research scientists and other problem solvers who have been shown to have ideas and begin to find solutions to problems at times and locations remote from those normally associated with the lab or other workplace (for examples, see Adams and Grimshaw, 2008). You should allow for a period of 'relaxed attention' that might last for a few days. During this time you should revisit the problem in a relaxed way; for example, when you are in a shower, on a bus or playing sport. Solutions to problems might suddenly occur to you in the bar or as you drift off to sleep, so always try to have with you a notebook and pen, or other recording device, to capture ideas that might appear at the most unexpected moments.

Table 1.2 Ideas in a box: novel approach for diagnosis of disease

ATTRIBUTES	SUB-ATTRIBUTES							
Applications	Medical	Veterinary	Agricultural	Military	Research	Bioterrorism	Forensics	Sport (athlete fitness etc.)
Location	Field	Hospital	Building site	Night club	In transit	School	Sports stadium	Space
Disease (including site affected)	Microbial infection	Cancer	Auto-immune	Cardiovascular	Genetic disorder	Food poisoning	Plant foliage	Plant roots
Host organism	Humans	Domesticated animals	Animals in sport	Wild animals	Farmed fish	Crops e.g. cereals, vineyards	Trees	Insects e.g. honeybees, silk worms
Techniques that might be exploited	Visual inspection	Culture of causative organism	Polymerase chain reaction	Colorimetric assay	Radiometric assay	Immuno-diagnostics	Biosensors	'Electronic nose'
Communicating information	LCD display	LED display	Sound	Smell	Taste	Vibration	Satellite	Mobile phone
Materials made from	Plastic	Metal	Glass	Paper	Organic	Wood	Living	Gaseous
Desirable properties	Inexpensive	Robust	Simple	Portable	Lightweight	Secure	Heat-resistant	Chemical-resistant
Added functionality	Computer	Online	Databases	Recyclable	Helpline	Work in tandem with other approaches/devices	Fully automated	International compliance

1.7   Working in groups – creative environments

Conventional teaching spaces such as formal seminar rooms or lecture theatres do little or nothing to foster creativity in their occupants. You will probably be aware that you are most creative in an environment that makes you feel comfortable and secure. During the last few years a number of universities, government agencies and other UK organisations have addressed this issue by introducing novel spaces that foster creativity (Figure 1.9). Once you and your classmates or colleagues have decided to have an idea-generation session you should, at the very least, ensure that the space you use allows face-to-face contact between participants (e.g. arrange seats in circle(s)

Figure 1.9   Creative environments (Images courtesy of: Dr Jeremy Rowntree, Department of Biochemistry, University of Oxford; InQbate CETL, University of Sussex; ALPS CETL, University of Leeds.)

rather than rows), provide lots of materials for recording results (flipcharts, marker pens, Post-it Notes etc.) and build-in regular comfort breaks with access to food and drinks.

1.7.1 Tutor notes

It is clearly very important to provide a physical space that promotes individual and group creativity. In addition, a creative environment/atmosphere should be one in which students feel comfortable and confident about expressing their ideas.

- Give *all* participants the opportunity to contribute ideas. The freely available website *Creativity in the Biosciences* (www.fbs.leeds.ac.uk/creativity; Case study, Section 1.5.1) is designed to promote individual creativity. If you wish, you can use this facility to encourage individuals to have and pursue ideas of their own prior to group work.

- Keep things informal and avoid rigid hierarchies. Avoid, at all costs, the classic lecture room configuration with chairs in neat rows, facing the front. Instead, wherever possible, encourage students to work in small, interactive groups with seating arranged e.g. in a circle.

- Allow the students breaks from time to time to keep them creative.

- Provide students with a great deal of encouragement. Focus on providing positive feedback and avoid negative criticism. Welcome even the most outlandish of ideas; these can be valuable because they often lead to really novel suggestions and original solutions to problems. They can also inject a very welcome element of humour to the proceedings!

- It can be useful to set goals as this can instil a sense of purpose; but, ensure the goals are realistic and that the students have time to achieve them.

If you are fortunate enough to be involved in the design of a *new* creative space within your institution, then you should give some thought to how you can foster the interdisciplinary interactions that can be so important for creativity (see also 'Make connections, Section 1.4).

Gerald Rubin is Director of Janelia Farm, the new research campus of the Howard Hughes Medical Institute. He describes how the design of Janelia Farm was influenced by other highly creative scientific institutions/environments including the MRC Laboratory of Molecular Biology in the UK, and the AT&T Bell Labs in the US (Rubin, 2006). Janelia Farm has therefore been created as a supportive environment that facilitates unscheduled interactions, and promotes collaboration between colleagues from a range of backgrounds who work together on interdisciplinary problems. Something of this nature has also been achieved in the new Biochemistry Department at Oxford University (Figure 1.9) which has been carefully designed to ensure colleagues must meet, and hopefully interact – note the grand piano close to a research laboratory!

1.8 Working in groups – facilitated creativity sessions

Clearly it's very important that you should be given every opportunity to achieve your own creative potential. For this reason, the approaches and techniques described until now focused on the promotion of individual creativity. However, interactive and well-managed *group* sessions can be extremely productive, leading to the generation of many novel ideas and solutions to problems.

The emphasis here is on *well-managed*. Unfortunately, group activity sessions frequently lack the expert facilitation that is essential if these techniques are to succeed. For example, most people appear to believe that traditional and often rowdy brainstorming sessions, with participants shouting out ideas that are recorded on flipcharts, invariably lead to the generation of lots of original and useful ideas. In practice this is not the case, and the results of a number of studies indicate that these traditional sessions, that are difficult to manage, are among the least effective approaches for the generation of ideas (Furnham, 2000). If you have ever participated in a brainstorming session then it's likely you are aware of one of the major problems associated with traditional flipchart brainstorming: some people talk too much! These extroverts may dominate the proceedings to the extent that shy, reticent group members feel intimidated and may be reluctant to put forward what may be highly creative suggestions and ideas. In this section we therefore describe creativity techniques that encourage full participation by all members of the group.

1.8.1   Non-oral brainstorming

Brainstorming sessions are not all bad! When facilitated properly they can generate a great deal of enthusiasm and collaborative energy in participants; the trick is to harness these positive elements and encourage all those present to fully engage with each exercise. This can be achieved using non-oral techniques like the 'Brainwriting 6-3-5' and 'Post-it Note' approaches that help ensure participation by all members of the group.

1.8.1.1   Brainwriting 6-3-5

This powerful technique, developed by Bernd Rohrbach, will allow you to work together with friends and colleagues in processing a number of ideas in parallel. Form a circle with five of your friends/classmates and decide on a problem you would like to solve. Write the problem at the top of each of **six** sheets of paper divided into a grid as illustrated in Table 1.3. Each group member should write down or sketch (see also 'Visualisation', Section 1.5.3.5) ideas relating to the problem in each of the top **three** boxes. After **five** minutes each member should pass the grid to the neighbour on their right. The neighbour then uses the earlier ideas as a source of inspiration for their ideas. After five

Table 1.3   Pro forma for the Brainwriting 6-3-5 technique

Problem			
	Idea 1	**Idea 2**	**Idea 3**
1			
2			
3			
4			
5			
6			

minutes they should pass the paper to the next person, who will continue to build on the ideas. The process should be repeated until all of the group members have had the chance to contribute to all of the grids. You should then collect the grids and write the ideas on a whiteboard for consideration by the group as a whole.

1.8.1.2 Tutor notes

An alternative approach is to ask each participant to start by writing or sketching their three ideas in the top row of a sheet of paper but to then place the sheets in the middle of the group for others to work on in their own time. This will allow individuals to think at their own pace and may encourage the more reluctant students to participate more fully. Towards the end of the session collect the sheets, and collate and consider ideas as indicated above.

1.8.1.3 Post-it Note brainstorming

Another approach that will help ensure participation by all members of your group is 'Post-it Note' brainstorming. In this technique a problem is identified and each participant is given a pack of Post-it Notes. They are asked to write each of their ideas/solutions to the problem on a separate Post-it Note which they stick to a flipchart. The ideas are then considered by the group as a whole.

The ideas generated during brainstorming can be evaluated by the group at a later stage using the approach described in Section 1.10 or, alternatively, the Six Thinking Hats technique.

1.8.2 Six Thinking Hats

Edward de Bono's Six Thinking Hats technique (de Bono, 2000) helps avoid conflict and the adoption of entrenched, inflexible positions during problem solving and evaluation of ideas. It does this by ensuring that all members of the group focus in the same direction and think in parallel. There are six coloured hats (perhaps metaphorically, although it's much more fun to use actual hats!) and, with the exception of the group leader (see below) all members of the group wear the same colour of hat at the same time. Each colour is associated with a particular way of thinking and this forces everyone to think in the same way, as follows:

- **White hat:** focus on the facts and analyse idea(s); ask for more information or data if needed.

- **Red hat:** represents emotions, feelings, intuition; for example, you might feel threatened or excited by the idea; it's important to get these issues out in the open.

- **Black hat:** represents caution, criticism, pessimism; look for faults in the idea (even if it came from you!).

- **Yellow hat:** look for anything good about the idea; think positively and optimistically.

- **Green hat:** the hat of creativity; generate novel ideas (using the wide range of techniques described in this chapter and elsewhere) or think about how an existing idea can be adapted or improved.

- **Blue hat:** review the process and decide whether you are making progress; this is also the hat worn by the chairperson throughout the session – he or she may direct the group to wear one of the other hats e.g. the green hat when more ideas are needed.

Exercise

Use the Six Hats approach to deal with a problem that you and your classmates would like to solve. For example, your group may be about to embark on a final-year team project and must choose from a bewildering range of topics that might be addressed in laboratory, field, library, local commerce etc. Refer to Chapter 6 for advice on the issues you should consider when choosing a research project, then start things off by donning a blue hat and acting as group leader. Ask the other members of the group to wear the range of different coloured hats, always ensuring everyone (except you – you wear the blue hat throughout) is wearing the same colour of hat at the same time. This will allow the group to consider all of the issues from a range of perspectives and to reach consensus on the project the group should adopt.

1.9 How many uses for an old CD?

Exercise

Try using the Creativity Techniques for Individuals and Groups (Sections 1.4–1.8) to help you and your friends and colleagues come up with ideas for recycling old CDs. You can compare your results with ideas suggested by others – see Section 1.10.

1.10 Evaluating your ideas

You and your friends may find you have a long list of ideas and you will want to converge on only one, or perhaps a small number, to take forward. You can do this by devising a set of criteria and a scoring system that can be used to judge each idea.

For example, you may decide you would like to set up a business based on a novel use for CDs (Section 1.9). Some of the potential uses you may have come up with include: table coaster, pizza cutter, part of a reflective 'disco dome', bird scarer, part of a toy 'balloon hovercraft' (Figure 1.10), earring, clock face, Frisbee. In Table 1.4, a group member has scored the potential uses against the criteria *Originality*, *Usefulness/value*, *Safety* and *Ease of construction*.

Table 1.4 Evaluation of creative uses for CDs

Idea/application	Originality	Usefulness/ value	Safety	Ease of construction	Total
Table coaster	1	3	4	3	11
Pizza cutter	2	2	2	4	10
Disco dome	3	2	2	1	8
Bird scarer	3	4	2	2	11
Balloon hovercraft	4	3	2	1	10
Earring	2	1	1	2	6
Clock face	3	2	4	1	10
Frisbee	1	2	1	4	8

SCORE: 1– low; 4 – high.

Figure 1.10   Balloon hovercraft; a creative use for an old CD

Each member of the group should be asked to complete a grid and you can then determine the mean total score for each application. In the event of a tie, decide which of the criteria is most important and place increased weighting on that criterion. For example, you may feel it is essential that any new use for a CD is very safe. You should therefore double the scores for this criterion. This may help identify a winning use for the CD. If not, repeat with another criterion, and so on.

1.11   Putting your ideas into action

It is likely that, during the course of your career, you will have many useful ideas. However, in common with most people, it is unfortunately equally likely that you will not make the most of these ideas and realise their full potential. Most of us tend to lack the time, inclination or confidence to act on our ideas and ensure that they are fully exploited. You should try to be a true innovator like Edison (Section 1.1), and in Chapters 11 and 12 you will learn how to protect, research, communicate and fund your ideas. This should ensure that they are of maximal benefit to you and the world at large.

1.12   How you can achieve your creative potential

- Be **curious** about the unusual, and welcome unexpected results – they may lead to the next scientific breakthrough!

- Don't be afraid to ask questions and **challenge assumptions**.

- Experiment with **'creativity techniques'** (described in this chapter and elsewhere) until you find an approach that works best for you.

- Look for **connections** and **analogies** between the apparently unrelated.

- Try solving problems by **thinking laterally**/'outside the box'.

- Allow time for **incubation** of problems and ideas.

- Take every opportunity to **collaborate** and **network** with people from different backgrounds.

- Learn how to protect, communicate and fund your ideas (see Chapters 11 and 12).

1.13   References

Adams, D.J. and Grimshaw, P. (2008) Creativity and innovation in the biosciences. In *Enterprise for Life Scientists: Developing Innovation and Entrepreneurship in the Biosciences*, Ed. Adams, D.J. and Sparrow, J.C. Bloxham: Scion, pp. 25–51.

Adams, D.J., Beniston, L.J. and Childs, P.R.N. (2009) Promoting creativity and innovation in biotechnology. *Trends in Biotechnology* **27**, 445–447.

de Bono, E. (2000) *Six Thinking Hats*. London: Penguin Books.

Furnham, A. (2000) The brainstorming myth. *Business Strategy Review* **11**, 21–28.

Kirton, M. (1976) Adaptors and innovators: a description and measure. *Journal of Applied Psychology* **61**, 622–629.

Michalko, M. (2006) *Thinkertoys. A Handbook of Creative-Thinking Techniques* (2nd Edition). Berkeley: Ten Speed Press.

Rubin, G.M. (2006) Janelia Farm: an experiment in scientific culture. *Cell* **125**, 209–212.

Sloane, P. (2009) *The Leader's Guide to Lateral Thinking Skills* (2nd Edition). London: Kogan Page.

Zwicky, F. (1969) *Discovery, Invention, Research through the Morphological Approach*. New York: Macmillan.

1.14   Additional resources

Bar-Cohen, Y. (2006) *Biomimetics: Biologically Inspired Technologies*. London: Taylor & Francis.

Buzan, T. and Buzan, B. (2007) *The Mind Map Book*. Harlow: BBC Active.

Byron, K. (2009) *The Creative Researcher: Tools and Techniques to Unleash your Creativity*. Cambridge: Vitae, Careers Research and Advisory Centre (CRAC) Ltd (www.vitae.ac.uk/CMS/files/upload/The_creative_researcher_Dec09.pdf). This guide provides practical information and advice about creativity in a research environment. Accessed April 2011.

Daupert, D. (1996) The Osborne-Parnes Creative Problem Solving Process Manual (www.kaplans.com/Daupert/Daupert%20Opening.htm). Accessed February 2011.

Forbes, P. (2005) *The Gecko's Foot. Bio-inspiration Engineered from Nature*. London: Fourth Estate.

Higgins, J.M. (2006) *101 Creative Problem Solving Techniques*. Florida: New Management Publishing Company.

Van Gundy, A.B. (2005) *101 Activities for Teaching Creativity and Problem Solving*. San Francisco: Pfeiffer.

The UK Centre for Bioscience, Higher Education Academy (www.bioscience.heacademy.ac.uk/) is a national organisation that provides excellent resources for students and academics. In particular, see the Centre's *Creativity Skills Short Guide* (available at www.bioscience.heacademy.ac.uk/resources/shortguides.aspx). Accessed February 2011.

The freely available *Creativity in the Biosciences* website (you will need to register) promotes creative approaches in individuals and groups. www.fbs.leeds.ac.uk/creativity/. Accessed February 2011.

A very useful website is www.mycoted.com/; dedicated to enhancement of creativity and innovation, it provides a repository of tools, techniques, exercises, puzzles etc. Accessed February 2011.

2 Problem solving – developing critical, evaluative and analytical thinking skills

Tina L. Overton

2.1 What is problem solving?

Whenever there is a gap between where you are now and where you want to be and you don't know how to find a way to cross that gap, you have a problem.

(Hayes, 1981)

Problem solving is what you do when you don't know what to do.

(Wheatley, 1984)

You may feel that the problem solving you have participated in until now, at school or university, does not necessarily fit these definitions. If a problem is familiar, and you know how to solve it to get the single correct answer, it could be argued that what you are doing is completing an exercise rather than solving a problem. In fact, with practice and familiarity, many problems become mere exercises. You will have noticed this already in your studies as you become more practiced at solving certain types of 'problems'. This shift is often described in terms of moving from being a novice problem solver to an expert problem solver.

The term 'problem solving' is used to describe a wide and varied range of activities. These activities by no means require the same set of skills and abilities to succeed. If we look at a couple of problems you will be able to see how different they are.

Problem 1

The pH of blood is 7.4. Calculate the concentration of hydronium ions, H_3O^+.

Effective Learning in the Life Sciences: How Students Can Achieve Their Full Potential, First Edition.
Edited by David J. Adams.
© 2011 John Wiley & Sons, Ltd. Published 2011 by John Wiley & Sons, Ltd.

Problem 2

Vitamin C is present in many foodstuffs. Assuming that synthetic vitamin C is unavailable, how many oranges would you need to eat to provide you with a sensible daily dose?

Let us analyse these two problems in terms of three simple variables: whether data are provided; whether the method required to solve the problem is well known; and whether the problem leads to a single correct answer.

In Problem 1 you are given all the data that you need. You have probably solved many problems of this sort in the past so the method is familiar to you and there is a single correct answer to the problem. This is an *algorithmic* problem and typical of the problems that students tackle during an undergraduate degree, especially in the early stages.

In Problem 2 the data are not given to you. You would have to make some estimations or carry out some research in order to generate some sensible data. You would have to decide how to go about tackling the problem as it may not be like problems you have seen before. And there certainly won't be a single correct solution to the problem, but a range of sensible outcomes. This sort of problem is defined as an *open-ended* problem and is very different from the algorithmic type. Open-ended problems are most often encountered during group activities, laboratory work and final-year projects (see also Chapter 6).

2.1.1   *Tutor notes*

You may like to try using problems of this type with two sets of rules. The first would ban the use of calculators or research (or Google), forcing students to make estimations and rough calculations. The second set of rules would allow the use of calculators and research, enabling students to come up with more 'accurate' answers. The students are likely to tackle the problems in quite different ways under the two sets of conditions.

You might expect to be faced with a wide range of problem solving activities during your degree studies, and these are likely to extend from the algorithmic through to the most open-ended problem. It is worth noting that being adept at solving algorithmic problems does not necessarily mean that you will be good at solving open-ended problems, and vice versa.

It is very difficult to teach anyone to solve problems. There are some strategies that you can employ to help you and some skills that will undoubtedly be of value. The aim of this chapter is to introduce some of these skills and strategies. But, in the end, it is practice and experience that improves problem-solving skills.

2.2   Problem-solving strategies

Over the years many models of problem solving have been proposed. Some of the numerous strategies or techniques are more useful for bioscientists than others. Among the most relevant are the following.

Analogy: using a solution that solved an analogous problem. This is particularly useful when solving algorithmic problems. You may often find yourself looking back through previous problems to find something similar. For example, you could use the solution to Problem 1 (above) to help you solve the following problem:

Problem 3

The pH of saliva is 6.8. Calculate the concentration of hydronium ions, H_3O^+.

Research: adapting existing solutions to similar problems. You may have used this technique before when you tried to find examples of similar problems and adapted strategies previously used by others.

For example, you could start from the solution to Problem 1 (above) to help you solve the following:

Problem 4

The hydronium ion concentration in a sample of saliva is 8.00×10^{-7} M. Predict whether its pH lies within the normal range.

Hypothesis testing: assume a possible solution to a problem then try to prove or disprove it. You may have used this technique before when working backwards through a problem from the known correct answer. This technique is useful only when an answer is known.

Divide and conquer: break down a large, complex problem into smaller, more solvable or familiar problems. This is a useful strategy but it takes practice and experience to spot the smaller elements that make up a complex problem.

Here is a simple example. Problem 5 can easily be divided into simpler steps, as shown in Problem 6.

Problem 5

Blood is an example of a buffered solution. Human blood is slightly basic and has a pH of approximately 7.40. If the pH falls, a condition known as acidosis can occur. Death may arise if the pH drops below 6.80. How many times greater is the hydronium ion concentration at pH 6.80 than at pH 7.40?

Problem 6

Blood is an example of a buffered solution. Human blood is slightly basic and has a pH of approximately 7.40. If the pH falls, a condition known as acidosis can occur. Death may arise if the pH drops below 6.80.

What is a buffered solution?

Calculate the hydronium ion concentration of normal human blood at pH 7.40.

Calculate the hydronium ion concentration of human blood at pH 6.80.

Calculate how many times greater the value at pH 6.80 is than at pH 7.40.

Lateral thinking: this approach involves developing creative solutions and coming at the problem from an oblique or nonlinear angle. Some of the techniques covered in Chapter 1 fall into this category and are useful for problem solving, especially for problems of the open-ended type.

Problem 7

How many oxygen atoms are there in the room you are in?

There are several different ways of thinking about this question. If you want to start by measuring the dimensions of the room, calculating its volume and hence the volume occupied by oxygen, taking account of the percentage of oxygen in air and then converting this volume to moles, you are taking a rational approach. However, consider that the question asked for the number of oxygen *atoms*. Perhaps there are no oxygen atoms in the room? Or are there twice as many oxygen atoms as there are O_2 molecules? You might also consider that you are sitting in the room. How many oxygen atoms do you contain? And the furniture, come to that? Given these considerations, do you still consider the number of oxygen atoms in the air to be significant?

Means-end analysis: taking small steps that move incrementally toward a solution. This is similar to divide and conquer and may involve several attempts and iterations before a sensible answer is approached.

Brainstorming: see Chapter 1. You may use this approach when starting out on a group project or assignment. It is an effective way of generating lots of ideas. See also Chapter 6 (Research projects).

Trial and error: trying out many possible solutions until one is found. You may have used this strategy when all other attempts have failed.

Some models of problem solving identify a number of stages in the process. One of the earliest of these was the method proposed by Polya in 1945 that involves four steps:

Understand the problem.

Devise a plan.

Carry out the plan.

Look back and review.

This is a simplistic model and you will be wishing that problem solving in your bioscience course was as straightforward as this. If we unpack this simple structure a little more then we find activities which may help move us towards a solution.

Understand the problem:

Restate the problem in your own words.

Draw a picture or a diagram that might help you understand the problem.

Do you have all the information to enable you to find a solution?

Devise a plan:

Make a list.

Eliminate possibilities.

Use reasoning and logic.

Develop, find or solve an equation.

Look for patterns.

Solve a simpler problem.

Use a model.

Work backwards.

Carry out the plan.

Look back and review.

More recently, George Bodner (2003) proposed the following model based on his observations of how science students solve problems. It is a better reflection of how complex the process really is:

Read the problem.

Now read the problem again.

Write down relevant information.

Draw a picture, make a list, write equation etc.

Try something.

Try something else.

See where it gets you.

Read the problem again.

Try something else.

See where this gets you.

Test intermediate results.

Read the problem again.

Strike your head and swear.

Write down 'an answer'.

Test to see if answer makes sense.

Start again or celebrate.

This long and frustrating process may look familiar to you.

A model that is based on the principles of the Scientific Method and used extensively in industry is the PDCA or Plan–Do–Check–Act method. The Scientific Method is described in more detail in Chapter 6, but forming and testing hypotheses is at the heart of it. The PDCA approach involves establishing what is required from the problem and planning a strategy, carrying out that strategy, checking or reflecting on the results and acting on the outcomes. The process is iterative in that the cycle is repeated and knowledge accumulates with each cycle. This method can be particularly useful during project work, or when managing a group assignment during problem-based learning as illustrated in Section 2.2.2.

2.2.1 Tutor notes

There are lots of websites that contain lateral- and logical thinking puzzles. Two of the best are www.fojl.com and www.rinkworks.com/brainfood/c/logic.shtml. It can be fun and beneficial to give students non-bioscience puzzles as a warm up to a problem-solving session.

2.2.2 Case study

You could use the PDCA method to manage a group assignment during problem-based learning. PDCA is based on a cycle:

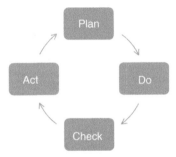

The cycle could be easily adapted to provide a structure to your group assignment, as shown below.

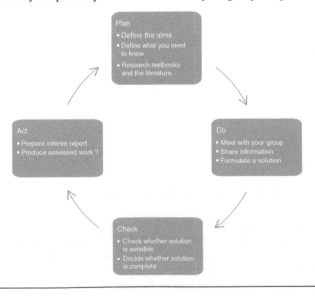

Most of the models of problem solving have some key steps in common: reading the problem carefully, making a representation of the problem, trying out a strategy, evaluating the solution, trying again and so on. Key to these steps is the ability to think critically, which enables us to evaluate a problem, define what needs to be done, review the quality of a solution, and recognise whether or not the solution is appropriate. We will now focus on critical thinking and other analytical skills which will enable you to become a more effective problem solver.

2.3 Critical thinking

The ability to think critically is one of the key, desirable qualities of a graduate and has been described in various ways. One of the more simple definitions was given as long ago as 1941 by Edward Glasner and describes critical thinking as:

(1) an attitude of being disposed to consider in a thoughtful way the problems and subjects that come within the range of one's experiences; (2) knowledge of the methods of logical inquiry and reasoning; and (3) some skill in applying those methods.

For scientists, critical thinking involves the following abilities:

* formulate strategies to solve unfamiliar problems;

* gather and use relevant information and data;

* recognise underlying assumptions and values;

* evaluate evidence and arguments;

* recognise logical relationships;

* test conclusions and generalisations;

* use judgement in decision making.

These skills could also be described as scientific thinking. They develop with experience and practice, but will not be acquired if learning fails to stretch us intellectually. So, problem solving of a complex type will help develop these skills. There are exercises which are designed specifically to develop critical thinking skills, and some are now being integrated into pre-university-level courses. You may therefore already have experience of these activities.

2.3.1 Tutor notes

If you'd like to help your students to develop critical thinking skills further, you might like to look on the website for the TSA. TSA (Thinking Skills Assessment) is a pre-interview admissions test for some of the University of Oxford's undergraduate courses. There are examples of the tests online which are designed not to need any underlying knowledge. Some may be more relevant than others to the biosciences, but all provide opportunities to develop thinking skills. For examples of how these styles of questions have been developed for use in undergraduate programmes, see Garratt *et al.*, 1999. This book is full of problems designed specifically to develop critical thinking skills. Although designed for chemistry there are problems relevant to the biosciences.

2.3.2 Case study

The information for this case study is taken from the freely available Critical Thinking site: www. criticalthinking.org.uk/ that is intended primarily for A-level students. In the following example, the evidence in a report is reflected upon and both the positive and negative sides of the article are presented. Too often the public accept sensationalist news articles at face value. The development of critical thinking skills will enable you to evaluate arguments and see the bigger picture.

Read the report on the BBC website of a doping scandal on the Tour de France in 2008, http:// news.bbc.co.uk/sport1/hi/other_sports/cycling/7655977.stm.

The fight against drugs cheats in sport goes on. One of the sports where doping has been widespread is cycling. Today, another two riders were found to have used banned substances to gain an advantage in this year's Tour de France, taking the total to six.

It's always difficult to know what to think when news of a failed drug test comes through. It's possible to put a positive spin on it, as the German cycling federation (BDR) chief has done:

'It is a shock, but it is also good news,' said BDR president Rudolf Scharping. 'The ever tighter net of the anti-doping investigators is making sure that practically no-one is getting through anymore.'

As Scharping says, one explanation for an increase in failed drug tests is that the tests are getting better. If we take this view, then although a failed drug test may be bad news in the short term, at least it can reassure us that we have effective tests and give us hope that the sport will soon be clean.

An alternative explanation for an increase in failed drug tests, though, is that there has been an increase in drug use. Our tests may be no more effective than before (the tests may have improved, but the methods used to evade detection may have improved too), and the reason that we're catching more riders may be that more riders are taking banned substances. In that case, each failed drug test just shows that cheating is prevalent.

There does seem to be a new determination to clean up cycling, but despite what Scharping says it isn't obvious that drug tests coming back positive shows that we're succeeding.

2.4 Critical reading

The key to getting started on a problem is deciphering what is being asked: identifying what it is you are required to do. Tutors often use authentic or contemporary contexts to make problems more interesting. Using a real-life context certainly does make problems more engaging and motivating, but an interesting context may also act as a barrier between you and the nub of the problem. Pulling out relevant information for a complex problem takes practice. You need to develop the skill of critical reading which will allow you to pick out the relevant information; it is a skill that you will find useful in a wide range of settings. Students are often faced with complex sources of information such as research papers, textbooks etc. and need to be able to extract the relevant facts. Have a go with the following examples.

2.4.1   Worked example   Rabbit calicivirus

(taken from Garratt *et al.*, 1999)

Rabbit calicivirus appeared in Europe in the mid 1980's, and kills over 90% of infected rabbits in the early stages of an outbreak. Because rabbits are a major pest in Australia, a research programme was established to test the effectiveness of the virus as a way of controlling the rabbit plague. The programme was established on an island, but the virus escaped to the mainland in October 1995.

Smith, Matson, Cubitt and others say that to justify releasing the virus in the first place, the Australian government should have first obtained clear proof that it infects just one species, the rabbit. AAHL researchers claim to have done just that. Between 1991 and 1996, they exposed 31 species of native and domestic animal to the virus. Samples from two New Zealand species that had been exposed to the virus were sent to the AAHL for analysis. The researchers measured the amount of antibodies and virus in the blood and organs of these animals. They also looked for any signs of sickness. According to GSIRO's public statements, those tests showed that the virus did not replicate or cause disease in any test animal. "Our testing of rabbit calicivirus is the most comprehensive study that we know of into the host range of an animal virus", says Murray. (Anderson A and Nowak R, New Scientist, 1997, 22 February.)

Consider the following questions:

(1) Smith, Matson, Cubitt and others indicated that clear proof should have been obtained that the virus infected only rabbits. AAHL researchers suggested they had obtained proof based on a range of criteria and experimental results. Were the criteria they applied reasonable?

You need to consider how you could obtain clear proof that the virus infects the rabbit and no other species. This is a demanding requirement/criterion even if you interpret 'no other species' to mean 'no other species of mammal living in Australia'. In this case, you need to consider how you would decide whether it is ever safe to use a biological method for controlling pests.

(2) Do the tests outlined here provide clear proof that there is no danger from the rabbit virus to the species tested?

You need to consider what you would regard as 'clear proof' that the species tested were not affected. You may feel you cannot make a judgement without more information about (for example) how many animals of each species were exposed to the virus, how and over what period the animals were exposed to virus, or how long the period of incubation was.

(3) Compare the statement attributed to Murray with the claim attributed to the AAHL researchers; are they consistent?

You need to consider whether carrying out 'the most comprehensive study that we know of' can reasonably be regarded as obtaining 'clear proof that (the virus) infects just one species'. If you think the two statements are not consistent, you need to consider what you would regard as clear proof (see (1)).

2.5 Using judgement

When solving problems we have to use our judgement in order to, for example, decide whether an answer is sensible, identify a context for a problem or make some estimations of data. Using judgement takes practice and requires knowledge relevant to the context of the problem. You will frequently be required to use judgement in those situations where the first obvious answer might be 'it depends'. For example, if you were asked how you might analyse the levels of sugar in a sample of fruit juice, the answer may well depend upon what equipment you have available, how many samples you have to analyse, who is going to carry out the analysis, what the cost of various methods is, whether there are any safety implications, whether you are required to use a standard operating procedure etc. So there is not a single, easy answer and you would have to use your judgement to come up with a solution.

2.5.1 Worked example Oestrogens in the food chain

(taken from Garratt *et al.*, 1999)

Oestrogens promote growth in cattle, but are banned because of concerns over their entering the food chain. Farmers have been brought to court on suspicion of using these illegal drugs. In such cases an analyst may be required to state the probability of obtaining a false positive result (i.e. detecting trace quantities of illegal oestrogen in meat when none were actually present). In a particular case where oestrogen was apparently detected, the analyst claimed that the chance of obtaining a false positive was 1 in 5000.

Which of the following statements do you agree with?

(1) The analytical evidence establishes guilt beyond reasonable doubt.

(2) The analytical evidence establishes guilt on the balance of probability.

(3) On the basis of the analytical evidence alone, there is insufficient evidence to prosecute.

Do you think it would be reasonable to use the same criterion (i.e. probability of obtaining a false positive of 1 in 5000) in a case which involved:

(1) a butcher suing the farmer because he had been supplied with unsaleable meat due to the oestrogen contamination?

(2) banning by an athletic association for illegal drug use?

(3) prosecution for murder?

There are at least two issues to consider here. First, on what basis can the analyst claim to know that the probability of obtaining a false positive is 1 in 5000? You might suggest that it is based on testing a very large number of samples which are known to be free from oestrogen and finding that 1 in 5000 gave a positive result; it would be hard to justify this value if less than 100 000 samples had been measured, and this seems unlikely. Whatever method has been used, the value will be an estimate, and the precision of the estimate is likely to be unknown.

The second is whether someone should be convicted if there is a 1 in 5000 chance of a mistake; at what level of doubt do you give the benefit of the doubt? There is a roughly 1 in 5000 chance of tossing 122 successive heads (or tails) with an unbiased coin. If you tossed 122 successive heads would you conclude that you had a weighted coin?

Most people would feel that the more serious the charge the lower should be the probability of obtaining a wrong analytical result by chance. Against this, one should remember that a conviction rarely depends on a single analytical result, but on several pieces of evidence.

Estimating data when solving a problem also requires the use of judgement and underlying knowledge. For example, if you were asked to estimate the volume of the oceans (without access to any data or the internet!) you would have to make some sensible estimations of the surface area of the Earth, percentage covered by water and the average depth. Have a go at the following examples:

What volume of oxygen is produced by a spider plant in a year?

How many carbon atoms are there in a human being?

2.6 Constructing an argument

An important aspect of solving a problem, and of reflecting on and presenting the solution, is the ability to construct a coherent and logical argument. The ability to construct arguments is another aspect of critical thinking. An argument can be described as a connected series of statements leading to a preposition. The statements are linked to the preposition by words or phrases such as 'therefore', 'thus', 'it can be inferred that', and 'it follows that'. For example:

Bioscience is a fascinating subject

I get good grades in bioscience

are statements and can lead to the conclusion:

I study bioscience at university.

Beware of false logic though. For example, *I study bioscience at university; bioscience is a fascinating subject therefore I get good grades in bioscience*, does not necessarily follow.

2.6.1 Worked examples

Construct a logical argument from the following statements.

The efficiency of utilisation of visible solar radiation for photosynthesis is limited by factors such as the quantum efficiency of photosynthesis, reflection of light from leaves, photorespiration and lack of leaf cover at certain times of the year.

> *Only a small fraction of the solar radiation reaching the ground can be recovered as chemical energy in crops.*
>
> *The energy needed to support photosynthesis is supplied by the electromagnetic radiation in the visible region of the spectrum, and this makes up about half of the total solar radiation incident on the Earth's surface.*

A logical argument can be constructed by linking the statements together in this way:

> *The energy needed to support photosynthesis is supplied by the electromagnetic radiation in the visible region of the spectrum, and this makes up about half of the total solar radiation incident on the Earth's surface. The efficiency of utilisation of visible solar radiation for photosynthesis is limited by factors such as the quantum efficiency of photosynthesis, reflection of light from leaves, photorespiration and lack of leaf cover at certain times of the year. **Therefore**, only a small fraction of the solar radiation reaching the ground can be recovered as chemical energy in crops.*

In the following example there is the possibility of choosing a structure which demonstrates false logic of the sort sometimes encountered when students interpret results.

> *An acidic solution containing aspirin reacts with iron(III) to give a violet solution.*
>
> *Solution X reacts with iron(III) to give a violet solution.*
>
> *Solution X contains aspirin.*

The preferred sequence is:

> *An acidic solution containing aspirin reacts with iron(III) to give a violet solution. Solution X contains aspirin. Solution X reacts with iron(III) to give a violet solution.*

Beware of the temptation to give:

> *An acidic solution containing aspirin reacts with iron(III) to give a violet solution. Solution X reacts with iron(III) to give a violet solution. Solution X contains aspirin.*

This is an example of false logic. There may be compounds other than aspirin in Solution X (such as other phenols) that will react with iron(III) to give a violet solution.

2.7   Visualisation – making representations

Research evidence indicates that making representations can help in the problem solving process (visualisation can also help when you are trying to be creative – see Chapter 1). A representation can be as simple as a drawing representing part of the problem scenario, a molecular structure, a mind map (see Chapter 1), a table, graph or a doodle that portrays your thinking. Making representations is a very useful way to break the problem down into smaller, more manageable sections or to visualise a complex problem. However, it is not something that can be taught and, for a given problem, each person may make a different representation. It's a good idea to get into the habit of making representations, as this should certainly help develop your problem-solving abilities. It will help prevent your working memory from becoming overloaded, because putting things down on paper frees up memory space.

2.7.1 Case study

Imagine that you were asked to solve the following problem.

Many commercial hair-restorers claim to stimulate hair growth. If human hair is composed mainly of the protein α-keratin, estimate the rate of incorporation of amino acid units per follicle per second.

There are several sensible ways to solve this problem. One of them might involve visualising the growth of hair in a diagram:

It might then be useful to visualise an amino acid unit and estimate its length:

2.8 Other strategies

2.8.1 Working together

Depending on the context within which you have been asked to solve a problem, it may be appropriate to work with another student to reach a solution. In many ways individual problem-solving activities are fairly artificial as, in most types of employment, you will be expected to solve problems in teams and you will be able to consult more experienced colleagues. Solving a problem with other people has many practical benefits. All problems place a demand on your working memory space and it is easy to become overloaded with information. If this happens then you will find it impossible to make progress. Working with even one other person means that you have combined working memory at your disposal. In addition, some people are better than others at deciding what is required by the problem statement; that is, they are better at seeing the wood from the trees. It is therefore likely that a group of people will be able to do this most effectively. You will also learn by observing how others solve problems and this will accelerate the rate at which you move from being a novice to an expert problem solver.

2.8.2 Making use of worked examples

The trick to becoming a proficient problem solver is experience and practice. Worked examples given by your tutor or presented in a textbook are an invaluable source of practice. They also provide a source of 'problem solving by analogy' as you may be able to use them to get started on similar or related problems. A good strategy is to work through a worked example and then to have a go at a similar, but not identical, problem. A related strategy is to use a technique called 'fading'. In this technique you work through a worked example almost to the end and then finish it off yourself. In the next problem you leave the worked answer earlier to complete it yourself, and so on, thereby fading away from worked example to problem solving.

Use worked examples and end-of-chapter problems whenever possible. They are designed to enhance both your conceptual understanding and problem-solving skills.

2.8.3 Open-ended problems

In the more open-ended problem-solving activities that you will encounter during problem-based learning, group assignments or final year projects, you may be presented with very unfamiliar problems and may not know where to start. First, you should define the problem statement for yourself and then identify what you will need to know to get started: you may need background knowledge about the topic, or details of how to conduct a series of experiments. Starting a problem equipped with necessary information is vitally important, but focus is required if you are not to become overloaded.

2.8.3.1 Case study

Consider the following and identify the information you would need in order to solve the problem.

Flatulence from sheep, cows and other farm animals accounts for around 20% of global methane emissions. The gas is a potent source of global warming because, volume for volume, it traps 23 times as much heat as the more plentiful carbon dioxide. A single cow can produce about 600 litres of methane per day. Scientists are currently researching the use of foodstuffs and vaccines that reduce the amount of methane produced in cows' stomachs. If, however, the methane produced by cows could be captured and used, how many cows would you need to generate enough methane to heat a house in winter?

You might want to start by identifying how much energy is required. If so, you could decide on the volume of water that would be heated, during 24 hours, for a central heating system. You would need to know the specific heat capacity of water and the average temperature change needed to heat water from tap water temperature to central heating temperature. Alternatively, you could cheat and look at your gas or electricity bill! Once you have the total energy required, you need to find the enthalpy of combustion of methane. It is then a relatively simple calculation to find how much methane is needed and, hence, the number of cows you would need to keep in your back garden!

2.9 Pulling it together

Ultimately, the only way that we become more proficient at solving problems is with experience, which means practice. You will be presented with a range of different opportunities to solve problems within your course: in lectures, tutorials, workshops, seminars, in the laboratory and in the field. Problems will range from closed, algorithmic ones to open-ended, ill-defined ones, and ideally you need to become confident at tackling them all. Problems encountered in employment are often open-ended in nature but can contain algorithmic elements within them. You undoubtedly need to acquire critical and logical thinking skills, but creativity will also enable you to find novel and effective solutions.

Finally, here are a couple of problems for you to have a go at. They are open-ended, in that there isn't a single correct answer and you haven't been given all the required data, but they also involve some algorithmic elements.

Problem 8

Taxol (paclitaxel) is a drug that is extracted from the bark of yew trees and used to treat breast, ovarian, lung and other cancers. When 1 mg of Taxol is administered orally to a rat it is found to be an effective dose and is cleared from the blood in three hours. Assuming that you wanted to prepare Taxol from yew trees, how many daily oral doses could be isolated from the bark from one tree?

Problem 9

We shed our outer skin completely approximately every three weeks. What weight of skin do we lose during this time? What does this imply in terms of molecules of amino acid incorporation into the skin per second?

2.10 How you can achieve your potential as a problem solver

- Don't be put off by an unfamiliar-looking problem.

- Rephrase a problem in your own words so that you are clear what is being asked of you.

- Break down complex problems into smaller steps.

- Look for similarities with previous problems.

- Carry out research to find similar problems or missing data.

- Don't be afraid to make estimations, but use sensible criteria, rather than guessing.

- Ask yourself if your solution is sensible.

- Don't give up; if you don't succeed straight way, have another go.

2.11 References

Bodner, G. (2003) Problem solving: the difference between what we do and what we tell students to do. *University Chemistry Education* **7**, 37.

Garratt, J., Overton, T. and Threlfall, T. (1999) *A Question of Chemistry: Creative Problems for Critical Thinkers.* Harlow: Longman.

Glaser, E.M. (1941) *An Experiment in the Development of Critical Thinking.* Columbia: Columbia University.

Hayes, J.R. (1981) *The Complete Problem Solver.* Philadelphia: The Franklin Institute Press.

Polya, G. (1945) *How to Solve It.* Princeton: Princeton University Press.

Wheatley, G.H. (1984) *Problem solving in school mathematics. MEPS Technical Report 84.01,* Purdue University.

2.12   Additional resources

Critical Thinking: The Unofficial Guide to OCR A-Level Critical Thinking. This website supports the OCR A-level
  critical thinking course. It contains many useful examples along with general discussion of critical thinking.
  www.criticalthinking.org.uk/. Accessed April 2011.
Developing Problem Solving Skills in Bioscientists. This report, published by the UK Centre for Bioscience, Higher
  Education Academy, presents the outcomes of a workshop: Engaging Students: What's the Problem. The
  report describes different strategies for promoting analytical, critical and creative approaches to problem
  solving in UK bioscience students. www.bioscience.heacademy.ac.uk/resources/problemsolving/.
  Accessed April 2011.
Watanabe, K. (2009) *Problem Solving 101: A Simple Book for Smart People*. London: Vermilion. A good primer for
  problem solving techniques and practice.

3 In the laboratory

Pauline E. Millican and David J. Adams

3.1 Introduction

Practical experimentation and analysis are at the heart of scientific activity. Our current body of scientific knowledge has been amassed over millennia, from the work of scientists developing theories from experimental investigation and progressively modifying these theories in the light of further study. Most important in this progress has been the ability of scientists to appraise experimental data critically and to assess how they relate to existing knowledge. As a student of the biological sciences you will be encouraged to participate in this process, most likely within the laboratory but also, perhaps, on fieldtrips (see Chapter 4) and through computational modelling.

Before arriving at university, most students will have had little opportunity to undertake laboratory work on a regular basis, and consequently many find the prospect of undergraduate practical sessions rather daunting. While trying to understand the relevance of new theoretical information, students must become proficient in handling specialist equipment; they must collect appropriate data and need to be aware of the reliability, or otherwise, of their results. Producing a suitable experimental report, with the data presented in the correct format and with valid conclusions, can prove stressful for those who are unprepared. If students do not understand what is important when describing their results, they may be unable to draw appropriate conclusions from them.

This chapter aims to guide you through aspects of laboratory work that students often find most difficult, pointing out how you can develop a positive approach to the sessions and understand how to gain the maximum benefit from them. A great way to start cultivating this attitude is to pay attention to the key experiments that led to particular scientific discoveries in your field. With an appreciation of how past scientific methods and techniques have been used to develop our current perceptions you will find it easier to become more engaged with, and enthusiastic about, the practical topics you encounter.

To give you an overview of how to approach your practical work to gain maximum benefit, the UK Centre for Bioscience has published a Student Short Guide, *Making the Most of Practical Work*, which contains useful details. The guide can be downloaded at www.bioscience. heacademy.ac.uk/ftp/resources/shortguides/practicals.pdf. It was written for students by

Effective Learning in the Life Sciences: How Students Can Achieve Their Full Potential, First Edition.
Edited by David J. Adams.
© 2011 John Wiley & Sons, Ltd. Published 2011 by John Wiley & Sons, Ltd.

students and is well worth reading as it nicely complements the information and advice provided in this chapter.

3.1.1   Tutor notes

A recent survey of academic staff across several bioscience disciplines (Wilson *et al.*, 2008) identified six major ways in which practical classes should be of benefit to students:

- clarification/illustration of theoretical concepts;

- development of skills and competence in practical work;

- application of the Scientific Method in problem solving and experimental design;

- improvement in data recording and handling abilities;

- personal development – greater confidence, engagement, reflection;

- improvement in understanding and implementation of appropriate safety measures.

The results of other surveys indicate that many potential employers consider it essential that bioscience graduates should acquire competence in practical laboratory skills (Archer and Davison, 2008; Saunders and Zuzel, 2010).

3.2   The Scientific Method

Progress in scientific knowledge relies upon experimental observation being developed into testable hypotheses which may be modified in the light of new observations. The results of well-designed experiments may support or refute a hypothesis but, either way, new information is gained. Where experimental results consistently conflict with an established theory it is important to design new experiments that probe further into the reasons for the discrepancy and to use the new data to modify the model. Unexpected results can lead to new insights which, in turn, may develop into exciting new research fields. More details about the strategies followed when applying the Scientific Method can be found in Chapter 6. In addition, the following case study illustrates the application of the Scientific Method in the field of microbiology.

3.2.1   Case study   Koch's postulates and the Scientific Method

Imagine that several individuals in an isolated community display signs of a devastating disease. The disease is thought to be caused by a microorganism but no one has been able to prove this. You are a microbiologist who is asked to establish whether or not there is a link between the disease and a particular bacterium. How would you proceed?

You should, of course, apply the Scientific Method (see Chapter 6), and the first thing to do is establish null and alternative hypotheses. In this case the null hypothesis is simply that there is no relationship between the disease and the bacterium; the alternative hypothesis is that the bacterium is the causative agent of the disease.

Now you must plan an investigation and carry out research, and at this stage you will find the pioneering approach of the nineteenth century microbiologist Robert Koch invaluable. Koch established criteria for the identification of microbial agents of disease, and these criteria are known as 'Koch's postulates'. Koch used these to identify *Mycobacterium tuberculosis* as the cause of TB, and *Bacillus anthracis* as the cause of anthrax. The postulates can be summarised as follows:

- The microorganisms must be found in abundance in all organisms suffering from the disease but should not be found in healthy organisms.

- The microorganism must be isolated from a diseased organism and grown in pure culture.

- The cultured microorganism should cause disease when introduced into a healthy organism.

- The microorganism must be re-isolated from the inoculated, diseased experimental host and identified as being identical to the original causative agent.

If you were to apply Koch's postulates to the devastating disease you have been asked to investigate, then you would expect to isolate the suspected causative agent of the disease from all affected individuals. You would then use your skills as a microbiologist to prepare a pure culture of the bacterium and use this to inoculate healthy laboratory animals (e.g. rats) that should all exhibit signs of the disease. You would then seek to re-isolate and purify the bacterium from the diseased host organisms.

Clearly Koch's postulates will provide a valuable framework for your scientific investigation. Following their application you will be able to gather data, analyse your results and decide whether or not they are consistent with the null or alternative hypothesis.

However, Koch's postulates also illustrate some of the complexities and issues associated with scientific research, as their use is not appropriate for all diseases caused by microbial pathogens. Can you think of problems that microbiologists may encounter as they apply Koch's postulates in the laboratory?

See the end of the chapter for a list of **some** of the problems associated with Koch's postulates.

Unfortunately, students of the biosciences are rarely introduced to the Scientific Method during introductory laboratory classes. Instead, the experiments they are asked to perform are frequently designed to illustrate specific, well-tested principles. This can lead to students holding preconceived assumptions about the outcome of experiments, and dismissing results that do not conform to their expectations. They may even be tempted to fabricate results. Clearly, this is precisely what students of the biosciences should **not** do: such an approach is both pointless and unethical (see also Chapter 9 for a more detailed consideration of the ethics of scientific investigation).

One of the most important skills you must acquire when undertaking laboratory work is the ability to reflect critically on the results of your experiments. If your data do not conform to an anticipated result, you should try to understand why. Poor experimental technique, problems with the materials or equipment, or a true deviation from the expectations could all be the cause. By developing a critical awareness of your own practical technique and an ability to assess the reliability of your results, you will quickly develop a more focused and mature approach to scientific experimentation. More advice on the processing and analysis of experimental data is provided later in this chapter.

3.2.2   Tutor notes

It helps to make plain to students that when their practical reports are assessed they will be rewarded for the discussion of all results, including apparently anomalous data; they should not aim simply to present a 'perfect' set of results. This will encourage proper reflection and honesty. Make clear to students what constitutes fabrication and cheating: they need to know that cherry picking or making up results is not acceptable. If students have a complete disaster with their results, suggest they get permission to use data from another student who they should acknowledge in their report. Dawson and Overfield (2006) reported that students were not always aware of what constituted plagiarism, so this must also be made clear (see also Chapter 9).

3.3   Preparing for a laboratory class

If you arrive at the teaching laboratory well prepared for the practical session, you will be more confident about what you are being asked to do and more focused and competent as you carry out the experiments. You are also likely to have a better idea of the sort of results you can expect and will therefore have an appreciation of whether or not the results you do obtain are sensible or unusual. This should lead to a much less stressful and much more satisfying laboratory experience.

3.3.1   Understanding the hazards

When undertaking laboratory work you must take very seriously the responsibility you have towards yourself and to others who use the laboratory, so you must behave appropriately. The most basic laboratory precautions include wearing appropriate protective clothing (fastened up) and keeping the working area tidy and free from clutter. Each set of practical instructions in your workbook will be accompanied by relevant safety notes that detail any hazards associated with the substances you will use and the procedures involved. It is vital that you read these and comply with any instructions given. If there is something you do not understand then you must ask for clarification.

3.3.2   Background reading

One of the most important things you can do to prepare yourself for a practical session is to read through the information about the practical in the laboratory manual **before** you arrive, and ensure you have a good grasp of the experiments you will be asked to perform. Make sure you understand the introductory material by reading lecture/tutorial notes and textbooks for clarification. You may find it useful to highlight or underline important details in the lab manual. If you make a note of any points you don't understand, you can ask about these at the start of the session. It may also be appropriate to contact the organiser of the practical class to ask about specific points.

3.3.3   Pre-prepared materials

Try to arrive at the practical session with pre-prepared blank tables and formatted graphs (i.e. graphs with scaled axes already drawn) so you can use these to quickly and efficiently record your results. This will save time during the practical class and will also allow you to identify quickly any anomalous points, and therefore experiments that might need repeating, within the allocated practical time. This preparation will pay dividends because you will be able to ask for advice while the issue is still fresh in your mind. If you record your results neatly in pre-prepared tables and graphs, you may also find that you do not have to reconstruct these tables and figures when you present them for assessment, thus economising on your time. Make sure you take everything else you will need to the practical session, e.g. lab manual, notebook, pen, calculator etc.

3.3.4 Preparation in the lab

Prior to starting experiments, take a few minutes to become familiar with all the materials and apparatus provided. Make sure you know what everything is for, lay out all materials logically and ensure that apparatus is clearly labelled; for example, label test tubes appropriately to ensure you know what they contain.

3.3.5 Tutor notes

Laboratory time is usually at a premium when introducing students to new techniques and equipment, so the use of online training resources to provide virtual training before the sessions can be invaluable. By giving students unlimited access to the resource and including a compulsory, short, online quiz, that must be completed to a satisfactory level prior to the lab practical, the time required for demonstrating equipment can be substantially reduced; more time is available for students to gain hands-on experience, and a more comprehensive understanding of the advantages and limitations of the various methods of collecting data can be instilled. Development of this approach resulted in the successful application of online learning and assessment in the University of Bristol's 'Chemlabs' project (www.chemlabs.bris.ac.uk). Before chemistry undergraduates at Bristol are permitted to work in the laboratory, they are required to undertake online training and assessment, including 'virtual laboratory' experiments. A similar approach in microbiology practicals at the University of Leeds led to significant improvements in student performance and satisfaction. These and other cutting-edge approaches to laboratory class teaching were reviewed by Adams (2009).

However, with the increasing availability of online support it may be tempting to completely replace expensive and difficult practical sessions with virtual approaches. The payoff in terms of financial and time gains may not compensate for the diminished student learning experience, as highlighted by the results of a recent survey which indicate that students prefer access to the more traditional methods of laboratory teaching **in addition to** online support (Quinn *et al.*, 2009).

3.4 Laboratory notebooks

It is very important that you clearly and neatly record all data **as you work**, not as half-remembered thoughts written down at the end of the laboratory session. Although many students prefer using loose-leaf paper, it is far better to use an A4 hardback notebook, with the first page as an index, and to number all the pages. Loose-leaf paper has a nasty habit of getting lost, tatty or filed out of order. With a notebook, this is less likely to happen.

In a teaching or research laboratory it is common practice to record **all** relevant information about the method being followed, including the quantities of solutes used to make up solutions, volumes used in assays etc. This will allow you to repeat the procedure if you feel something has gone wrong or if the experiment has produced interesting results. It will also provide a valuable record to show to others who may want to follow, or modify, your procedure.

As an undergraduate, you may not be required to reproduce material, such as the Materials and Methods section from the laboratory manual, in your laboratory notebook. Be aware, however, that anything you do that deviates from the given instructions must be noted; it may help to explain any differences between your results and those obtained by others. More detailed advice on how to present a formal laboratory report is given later in this chapter.

As indicated earlier, preparation of blank graphs and tables, which you can use to record results, will save time and will help you develop the habit of recording data neatly. Remember to leave plenty of space for your data. Leaving too little room can mean that the data will 'spill over' and look untidy. If the data you collect does not fill a table completely, does it matter? It will still look neat. You should also leave gaps for results that you obtain later, such as those contained in chart readouts or photographs.

3.4.1   Tutor notes

You may wish to consider use of electronic laboratory books (Badge and Badge 2009).

3.5   Laboratory equipment

It is essential that you should fully understand how to use laboratory equipment, and you should always refer to the operating instructions. These ought to be readily available in your laboratory manual or elsewhere. If you cannot understand the instructions, ask for help. Hopefully, this will prevent you from making expensive mistakes.

You will be aware that there are, in fact, many biological sciences that range from agriculture through nanotechnology to zoology. Each discipline utilises specialised and often highly sophisticated items of equipment, and we are unable to consider all of these here. However, as an undergraduate enrolled on a bioscience degree programme, you are likely to encounter a number of standard pieces of apparatus and we consider issues associated with the use of some of these items of equipment in the pages that follow.

3.5.1   Tutor notes

It is **essential** to check that all the equipment works and all the necessary materials are ready before the start of the practical session. Students are easily discouraged if they have to struggle with faulty equipment, or realise late in the practical session that their equipment has been set up incorrectly. This is particularly important for students in the early phase of their Level 1 studies, when confidence issues abound. Be aware that students with disabilities may require assistance to operate some equipment. Moveable laboratory benches can be used for those in wheelchairs, but thought must be put into the use of other facilities such as sinks.

3.5.2   Dispensing liquids – automatic pipettes

One of the most common sources of error in laboratory work is the incorrect use of automatic pipettes: 'automatic' does not mean 'lacking the possibility of error'. Students sometimes take a lot

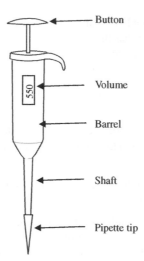

Figure 3.1   Automatic pipette set to deliver 550 μl

of convincing that their anomalous results are due to poor pipetting technique, and the problem is compounded when they make serial dilutions. Mistakes made at the beginning of a dilution series are magnified enormously throughout the series. After reading this section you will be more aware of the potential problems and know how to avoid them.

Common features for most automatic pipettes include a shaft onto which a disposable tip can be placed, a mechanism for adjusting the dispensing volume, a dispensing button and a two-stop system for collecting and then ejecting the fluid. Since there are several styles of automatic pipette, each operated in a slightly different manner, it is **very important** that, before using an unfamiliar style of pipette, you check how it works. A practical demonstration of how to use a pipette correctly has been created by the Genetics Department at the University of Leicester, and is available on the JORUM website at www2.le.ac.uk/departments/genetics/vgec/educators/post18/topics/recombinanttechniques/micropipette-1?searchterm=using%20a%20pipette.

3.5.2.1   Points to note when using an automatic pipette

- Use a pipette of the appropriate capacity. For example, use a pipette that can dispense up to 0.1 ml to dispense 0.1 ml. If you use a pipette that can dispense up to 1 ml for this manipulation, the amount dispensed will be delivered much less accurately.

- Attach the pipette tip securely to the barrel. A poorly fitting tip will allow air to be drawn into the pipette and prevent the correct volume from being dispensed.

- Set the dispensing volume mechanism to give the desired volume (if you don't know how to do this, ask).

- Press the button to the **first** stop position.

- Hold the pipette vertically and lower the tip into the fluid so that the lowest part of the tip is under the surface.

- Release the button **slowly** and steadily to draw the fluid into the tip. A steady release avoids turbulence and reduces the chance that fluid will shoot up the shaft of the pipette – this would prevent it from functioning accurately and might cause damage. Large pipettes can be fitted with filters which help to prevent this happening.

- Withdraw the pipette from the sample and carefully remove any excess liquid from the outside of the tip (avoid touching the orifice).

- To eject the fluid, place the tip against the side of the receiving vessel and press the button down steadily and slowly to the first stop, and then down to the second stop position to completely expel any remaining fluid.

- Keep the button depressed until you have withdrawn the tip from the receiving vessel (to prevent fluid being sucked back into the tip).

- Minimise any errors due to fluid clinging to the inner surface of a new tip by drawing up the correct volume of solution, ejecting it completely, and then drawing up the fluid for a second time before dispensing. This is particularly important when pipetting viscous fluids, such as blood, where a substantial amount of the fluid may remain on the inner surface of the pipette tip.

- Get into the habit of observing fluid as you draw it into the tip and subsequently expel it, to check you have pipetted correctly. With practice, you will be able to gauge whether the volume in the tip looks about right, so you will be able to spot immediately when there is a problem that will spoil your experiment.

3.5.3   Weighing solids – precision balances

It is important you choose a balance that is appropriate for the quantity you wish to weigh; otherwise you will incur unnecessary errors.

Portable balances are often called 'top pan' balances since the weighing pan sits on the top of the balance. These balances can detect, fairly inaccurately, a minimal weight of around 0.01 g (10 mg). They are therefore used routinely to weigh larger quantities of materials.

Analytical balances can be used to weigh quantities down to 0.0001 g (0.1 mg). Errors can easily be introduced when dealing with such small quantities, so extra precautions are needed to ensure that measurements are accurate.

3.5.3.1   Points to note when using balances

- The 'tare' facility allows you to subtract the weight of the weighing paper (or boat) from the total weight.

- Use an appropriately sized, clean spatula for the quantity to be weighed.

- Plastic weighing boats often develop static electricity which makes it difficult to transfer all of the weighed material into the receptacle. This can be avoided if you use a piece of weighing paper, folded down the middle (to make it easier to 'pour' the material into the receptacle) rather than a plastic boat. When weighing only a very small amount, use a smaller piece of weighing paper.

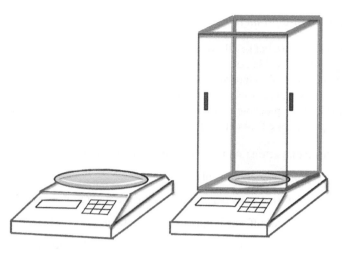

Figure 3.2   Top-pan and analytical balances

- Try not to cause draughts around the balance; the change in air pressure could influence the reading. This is especially important when dealing with analytical balances, so always close the doors to take your reading.

- Keep the balance clean and tidy, especially the pan, to avoid contamination by other substances (ask for a small brush to sweep away any debris).

- Clean your spatula carefully after use to avoid cross-contamination of substances.

3.5.4   *Measuring pH*

Many experiments must be performed at a specific and constant pH, and this is achieved using buffer solutions. A buffer compound (for example, sodium citrate, sodium acetate, 'Tris') is dissolved in water and the pH adjusted by the addition of an appropriate acid or base. To ensure that an appropriate amount of acid or base is added, the pH is constantly monitored using a pH meter.

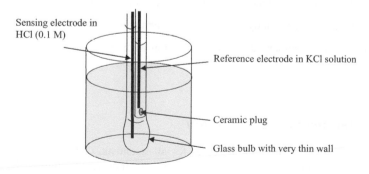

Sensing electrode in HCl (0.1 M)

Reference electrode in KCl solution

Ceramic plug

Glass bulb with very thin wall

Figure 3.3   Standard glass pH electrode

3.5.4.1 Calibration of a pH meter

Electrodes of pH meters must not be allowed to dry out and must therefore be stored in liquid when not in use. They must be calibrated before use, and calibration usually involves a two-point calibration procedure using two standard buffer solutions with pH values towards the extremes of the range of pH values for your experiments. The following is a general procedure for calibration of a pH electrode. However the procedure may vary between models produced by different manufacturers, and it is important to check details for each electrode before use.

(1) Adjust the temperature setting on the pH meter to room temperature.

(2) Switch the meter from 'standby' to 'pH'.

(3) Remove the electrode from the storage liquid and rinse it carefully using deionised water.

(4) Place it in the first standard buffer and adjust the meter until the readout shows the correct pH value.

(5) Rinse the electrode (to avoid contamination) and place it in the second standard buffer.

(6) Adjust the meter until it shows the correct pH for the second buffer.

(7) Rinse the electrode and place it in the solution whose pH you wish to measure.

3.5.4.2 How to adjust the pH of a solution

When adjusting the pH of a strong buffer solution it is common for students to overshoot the required pH. This can be avoided by adopting the following procedure. Add drops of concentrated acid or base to the buffer, with constant stirring, until the pH comes within 1–1.5 units of the required pH value. Now switch to using dilute acid or base ('bench' concentrations, for example 1 M acid or base) for the final adjustment. This change from concentrated to dilute acid/alkali at the later stages allows you to make the larger, initial adjustments quickly and then to fine-tune the pH to the required value. More dilute buffer solutions can be adjusted to the desired pH using only 'bench' concentrations of acids or bases.

3.5.4.3 Points to note when using pH meters

- Be gentle with the electrode; it is very fragile – keep it suspended safely in a clamp and **take liquids to it** for measurement.

- Ensure the ceramic plug of the electrode (Figure 3.3) is completely submerged in liquid and that the electrode does not come into contact with the walls of containers.

- Use a magnetic 'flea' to constantly stir a solution when adjusting the pH.

- Ensure the electrode is submerged in storage solution when not in use.

- Re-calibrate the electrode on a regular basis.

3.6 Calculations in the laboratory

Data collection and analysis are core activities for all scientists; therefore numerical skills are essential. However, it is very common for students starting their university course to lack confidence in this area because they have had insufficient practice in simple mathematical manipulations.

Particular confusion arises with the various degrees of scale and the relationship between different types of unit. These issues can easily be addressed, and this section aims to help you overcome any fears you may have about performing basic calculations so that you can approach them in a rational and systematic manner. You will find more help and advice, which should help build your confidence in the use of maths, in Chapter 7.

3.6.1 Units

In the biological sciences, mass (or weight), volume and time are commonly measured using standard units of grams, litres and minutes (or seconds), respectively. Often, however, you will be required to perform calculations with quantities on a much smaller scale. It is important that you understand the meaning of these different units, how they are represented in standard format and the values denoted by appropriate prefixes.

The units you will encounter most frequently in the biological sciences are emphasised in **bold** in Table 3.1. There are differences in convention across the scientific disciplines, and it is particularly worth noting that medical terminology often uses dl instead of 0.1 l; so it is important that biomedical students take extra care to avoid confusion over units.

From Table 3.1 you should be able to see that:

1 g is the same mass as	1×10^{-3} kg and	1 000 mg	
1 mg is the same mass as	1×10^{-3} g and	1 000 µg and	1 000 000 ng
1 l is the same volume as	1 000 ml		
1 ml is the same volume as	1×10^{-3} l and	1 000 µl	

Units of time: do remember that 1 minute 30 seconds is $1\frac{1}{2}$ (1.5) minutes and **not** 1.3 minutes, (similarly 1 minute 45 seconds is 1.75 **not** 1.45 minutes).

Note: remember to attach the correct units and their prefixes to any observation you make; this will allow you, and others, to understand what the numerical values mean. Bear in mind that some data (for example, light absorption measured by a spectrophotometer and some equilibrium constants) have no units because they represent a ratio of two numerical values with the same units.

Table 3.1 Units of mass and volume commonly used in the biological sciences

Name	Symbol	Numerical quantity	Scientific notation	Mass	Volume
kilo-	k	1 000	10^3	kilogram, kg	*Prefix not commonly used*
—	—	1	10^0	**gram, g**	**litre, l**
deci-	d	0.1	10^{-1}	*Prefix not commonly used*	dl *(more commonly used in medicine)*
centi-	c	0.01	10^{-2}	*Prefix not commonly used*	*Prefix not commonly used*
milli-	m	0.001	10^{-3}	**mg**	**ml** *(cm³ is more commonly used in chemistry and physics)*
micro-	µ	0.000001	10^{-6}	**µg**	**µl**
nano-	n	0.000000001	10^{-9}	**ng**	*Prefix not commonly used*
pico-	p	0.000000000001	10^{-12}	pg	*Prefix not commonly used*

3.6.2 Tutor notes

The usefulness of including the correct units alongside the relevant numerical values is often not apparent to students. It is worth encouraging students, from the start of their university career, always to include appropriate units, perhaps attributing a portion of their assessment marks for this aspect. Although dealing with units of time generally causes few problems for students, occasionally mistakes arise when data, collected in units of seconds, are not converted to units of minutes (or vice versa) for the final answer. Alternatively, students may convert the number of seconds into a fraction of a minute incorrectly.

3.6.3 Preparing solutions

In the laboratory you will need to prepare solutions containing precise amounts of substances (solutes) dissolved in fixed volumes of solvents (usually water, sometimes an organic solvent such as ethanol). You will weigh the substances in grams, or fractions of grams, and you will dissolve them in solvent at a specified concentration, for example, 0.5 g in 1 litre of water. However, concentrations in the bioscience laboratory are generally not expressed in grams per litre. More commonly, they are expressed in *moles* per litre, an approach that takes into account the markedly different masses of individual atoms and molecules. You will recall from GCSE studies that a mole is the amount of an element that contains 6.0221×10^{23} atoms. This is known as Avogadro's constant.

In the periodic table (Figure 3.4), the mass of a mole (or 6.0221×10^{23} atoms) of each element is indicated below its symbol. These are comparative masses (commonly called the relative molecular mass or M_r); they are based on a scale where the carbon 12 isotope has a mass of 12 units, and they are expressed in grams. So, for example, the mass of a mole of sodium (Na) is given as 22.99, and the mass of a mole of mercury (Hg) is 200.6. This means we have a value in grams for the mass of one mole of any element.

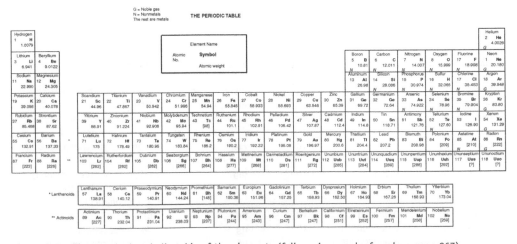

Figure 3.4 The standard periodic table of the elements (full version can be found on page 267)

By extension, the mass in grams of one mole of any molecule can be calculated by adding together the molar masses (M_r values) of all the individual constituent elements. For example:

A mole of sodium chloride has a mass of 22.99 g + 35.45 g = 58.44 g	A mole of glucose has a mass of $(12.011^a \times 6)$ g + (1.0079×12) g + (15.999×6) g = 180.15 g

[a]**Note:** in the above calculation, the mass of carbon is given as 12.011. This is slightly higher than the mass for carbon 12, due to small amounts of naturally occurring carbon 13 and 14 isotopes.

An important definition to stress is that:

One mole of a substance is its relative molecular mass (M_r) in grams.

The units of moles are **mol** (**not** m, which is the symbol for metres).

Since we know that 58.44 g is the mass of 1 mole of NaCl, it is relatively easy to see that if we want 2 moles we must weigh out twice the M_r value (i.e. 116.88 g), and if we want 0.1 mole we must weigh out 0.1 times the M_r value (i.e. 5.844 g). We are adjusting the value of the molar mass according to the number (or fraction of) moles that we require. This also applies to more awkward numbers, as illustrated in the table below.

Number of moles of NaCl required	Multiply the M_r by the number of moles required	Mass of NaCl to be weighed (g)
2	2×58.44	116.88
1.25	1.25×58.44	73.05
0.75	0.75×58.44	43.83
0.15	0.15×58.44	8.766

3.6.3.1   Why we use the mole as a unit

The reason scientists tend to quantify the amounts of substances in moles rather than mass is because chemical reactions occur between discrete numbers of molecules and we are interested in the *relative* proportions of whole molecules involved in a reaction rather than their actual masses. The complete reaction between glucose and oxygen in respiration illustrates this point:

$$6 \, O_2 + C_6H_{12}O_6 \longrightarrow 6 \, CO_2 + 6 \, H_2O$$

It is easy to remember that six moles of oxygen are needed to react completely with one mole of glucose to produce six moles of CO_2 and six moles of H_2O. In contrast, the numerical values of the masses turn out to be rather awkward: (191.988, 180.15, 264.054 and 108.089, respectively) and very easy to forget.

3.6.3.2   Preparing molar solutions

We define a one molar solution as:

One mole of a substance dissolved in a final volume of one litre of solvent (usually water).

This concentration represents 1 mole in one litre (1 mole/l; more correctly $1\,\mathrm{mol\,l^{-1}}$). It is normally expressed as **1 M** (*capital* M). **Do not confuse mol and M. The first is an** *amount* **and the second is a** *concentration*.

One way of making a 1 M solution is to weigh out the M_r (in g) of the substance and dissolve it in 1 litre of the solvent. So, to make 1 litre of a 1 M solution of NaCl you must dissolve one mole, i.e. 58.44 g, in water to a final volume of 1 litre.

Note: if one litre of water was added to 58.44 g NaCl, the final volume would be *greater* than one litre (since the NaCl also has a volume). So, if you were asked to prepare one litre of a molar solution of NaCl, the correct way to do this would be to dissolve 58.44 g NaCl in a smaller volume of water (e.g. 700 ml), then add water until you have a final volume of one litre. This may be done fairly crudely using a measuring cylinder; however, a more accurate (and preferable) method would be to use a volumetric flask.

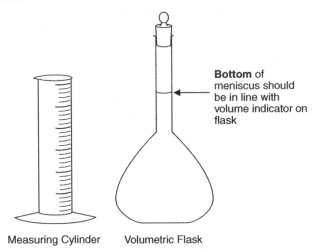

Bottom of meniscus should be in line with volume indicator on flask

Measuring Cylinder   Volumetric Flask

We have defined a 1 molar solution (1 M) as 1 mole in 1 litre. It follows from this that a 2 molar solution (2 M) is 2 moles in 1 litre; a 0.25 molar solution (0.25 M) is 0.25 moles in 1 litre; a 0.015 molar solution (0.015 M) is 0.015 moles in 1 litre; and so on.

3.6.3.3   Preparing solutions in volumes that are not 1 litre

In the laboratory it is unlikely that you will always have to prepare one litre of a solution, so this section explains how to prepare solutions of different volumes.

First, we shall consider how to prepare different volumes of a 1 M NaCl solution. Since the concentration must remain at 1 mole for every litre, *regardless* of the volume, this means that for a 2 litre solution you would need twice as much NaCl (i.e. 116.88 g) and for 1.5 litres of a 1 M solution you would require 1.5 moles of NaCl (i.e. 87.66 g). Thus, you are increasing the number of moles weighed out in proportion to the extra volume so that the concentration does not change.

In a similar manner, if you want to prepare less than one litre, you must reduce the number of moles used in proportion to the reduction in volume. **To do this correctly, it is important to always express the volume in *litres*,** which will give you the number (or fraction) of moles that you need. You then adjust the molecular mass by the same proportion (by multiplying the M_r by the number of moles needed in the required volume) to obtain the amount you must weigh out. This is illustrated in the table below.

Volume of 1 M solution required	Volume expressed in litres	Number of moles required for a 1 M solution	Amount of NaCl needed for the 1 M solution
1.75 l	1.75	1.75	$1.75 \times 58.44\,g = 102.27\,g$
300 ml	0.3	0.3	$0.3 \times 58.44\,g = 17.532\,g$
75 ml	0.075	0.075	$0.075 \times 58.44\,g = 4.383\,g$
1.65 ml	0.00165	0.00165	$0.00165 \times 58.44\,g = 0.096\,g$

3.6.3.4 Preparing more complicated solutions

In the bioscience laboratory, you are likely to be asked to prepare much smaller volumes of solutions containing molecules that are much larger than NaCl, and often at fairly dilute concentrations. For example, you might be asked to prepare 5 ml of a 25 mM solution of glucose. The best way to approach this problem is to start by considering the mass of solute needed to make 1 litre of a 1 molar solution and then make **two** adjustments to this value: the first to take account of the different concentration required and the second to take account of the different volume. The example below illustrates this process.

Worked example 1

You want to prepare 5 ml of a 25 mM glucose solution.

You know that the molecular mass of glucose is 180.15 g, so 1 l of a 1 M solution will contain 180.15 g glucose.

First, adjust this mass to give 1 litre of the required concentration:

You need a 25 mM solution; this is 0.025 moles per litre

0.025 moles has a mass of $(0.025 \times 180.15\,g) = 4.5\,g$

So 1 litre of a 25 mM solution contains 4.5 g glucose.

Secondly adjust this mass (i.e. 4.5 g) to account for the smaller volume:

5 ml is 0.005 l (*remember always to express the volume in litres*)

4.5 g in 1 litre has the same concentration as $(4.5\,g \times 0.005)$ in 0.005 litre, i.e. 0.0225 g (*you have scaled the 4.5 g down to the same extent as the volume*).

So 0.005 l (5 ml) of a 0.025 (25 mM) solution will contain 0.0225 g of glucose.

This is the same as 22.5 mg of glucose.

This example illustrates the importance of understanding units and having the ability to readily convert M to mM, mM to µM, g to mg etc.

To weigh out a small amount like 22.5 mg you would use an analytical balance rather than the cruder 'top pan' balance used for larger amounts (Figure 3.2). The small amount of liquid required for this solution can be measured out using a 'micro-pipettor' (Figure 3.1).

3.6.3.5   Other ways of expressing concentration

Sometimes it is not convenient to prepare a solution based on the number of moles present. For example, you may be asked to prepare a solution of a complex substance, such as bovine serum albumen or milk powder, that doesn't have a precise molecular mass. The concentrations of solutions of these materials are usually expressed as, for example:

5 grams of milk powder dissolved in 1 litre $= 5$ g/l or 5 g l^{-1}

You should familiarise yourself with degrees of scale so that it becomes second nature to realise that 1 g l^{-1}, 1 mg ml^{-1} and 1 µg µl^{-1} all represent the same concentration (compare units in Table 3.1).

Occasionally concentration is expressed as % w/v (weight per volume) and % v/v (volume per volume). The term % w/v simply refers to how many grams there are **in** 100 ml total solution (not how many grams of compound **plus** 100 ml solution). Normal saline is a solution whose concentration is most commonly expressed as 0.9% (w/v) NaCl, although it actually represents a 0.154 M NaCl solution. In the case of pure liquids such as ethanol, the term % v/v indicates how many ml of pure liquid there are in 100 ml of the total volume of the specified liquid and water.

3.6.3.6   Preparing dilute solutions from concentrated stocks

It is unlikely that you will leave university without having to prepare a range of serial dilutions or a set of standard samples from a more concentrated stock solution; for example, to create calibration data. Unless you are careful, it is very easy to make mistakes when calculating the relative proportions of diluent and stock solution required; so this section provides an explanation of how to get it right every time.

One of the simplest dilutions you may encounter is where you need to produce a 0.5 M solution from a 1 M stock solution. Since the required concentration is half that of the stock, it is relatively easy to see that you must mix together equal amounts of the stock solution and water.

When dealing with more awkward volumes, however, you will need to calculate the dilution factor first. This is the factor by which the original (stock) concentration is greater than the final concentration and represents the volume **into which** 1 ml of the stock solution must be diluted. (The diluting solution may not always be water but a buffer solution or an organic solvent.)

To calculate the dilution factor, you must divide the concentration of the stock solution by the concentration of the required solution. (**Note:** make sure that the concentration of stock and final concentration are expressed in the same units, M, mM etc.)

$$\text{Dilution factor} = \frac{\text{concentration of stock solution}}{\text{concentration of required solution}}$$

Applying this to the example above, we can calculate the dilution factor.

$$\text{Dilution factor} = \frac{1\,\text{M}}{0.5\,\text{M}} = 2$$

This means that you must add 1 ml of stock solution to water to make a total volume of 2 ml. Therefore, the amount of water required is also 1 ml.

If you want a volume larger than 2 ml, you must scale *both* volumes up accordingly. So if we wanted 50 ml of our solution rather than 2 ml we would need to use:

$$\frac{50 \text{ ml}}{2 \text{ ml}} = 25$$

25-fold more of each of the solutions is required, i.e. 25 ml of each.

The same reasoning applies to more complicated cases. This is illustrated in the following example.

Worked example 2

We want to use a 0.25 M stock solution of phosphate buffer to make 100 ml of a 5 mM solution.

First, we must express the stock solution and the final solution in the same units; in this case 0.25 M can also be written as 250 mM. We can then calculate the dilution factor:

$$\text{Dilution factor} = \frac{250 \text{ mM}}{5 \text{ mM}} = 50$$

This means that we need 1 ml of stock **in** 50 ml of water (i.e. 1 ml **plus** 49 ml water).
Since we want 100 ml not 50 ml, we must increase both volumes by:

$$\frac{100}{50} = 2$$

i.e. two-fold more of each of the solutions.

Therefore 2 ml of stock must be present **in** a total volume of 100 ml, i.e. 2 ml stock **plus** 98 ml of water.

This method also works for less-obvious cases. For example, to obtain a 50% solution from a 70% solution, the dilution factor is $70\% \div 50\% = 1.4$.

Therefore 1 ml of the stock must be diluted into 1.4 ml **total** volume.

This is the same as 1 ml of stock **plus** 0.4 ml of water (or a scaled up volume if required).

3.6.4 Tutor notes

Students often arrive at university without a complete understanding of how the mass of a compound is related to the number of moles present, or how to scale up or down from the M_r to obtain the mass required to prepare a specified solution. Especially prevalent seems to be the use by students of 'formulae triangles' into which they insert various terms with the expectation that (magically) the correct answer will appear at the end! It is strongly advisable to discourage this type of approach since, when students don't fully understand the calculation or the formula, they forget which way up the terms should go and make mistakes. By encouraging students always to start by considering the mass required to prepare a 1 M solution, and then to use the simple stepwise methods to adjust up or down from the M_r value (as illustrated in this chapter), they should feel more confident and in control of their ability to calculate the correct quantities to weigh out.

The preparation of diluted solutions also causes great difficulty to some students due to a lack of understanding of proportions and fractions. For help with preparing serial dilutions for microbiology classes, see: www.bmb.leeds.ac.uk/mbiology/ug/ugteach/bmsc1213/Micro/Micro_2/player.html.

3.7 Working in a group

Most students appreciate the opportunity to interact socially with other students and members of staff during practical work (Collis *et al.*, 2008). Usually, you will work closely with at least one other student during practical classes; occasionally you will be expected to work and cooperate as a member of a larger group. You should make the most of such opportunities, as learning to work successfully with others will bring great benefits. When everyone in the group is actively engaged in the work and contributing effectively, the practical is likely to progress smoothly and to finish well within the allotted time. Sharing responsibility, getting the work done and acquiring a good set of results makes the process of experimentation and the appraisal of the data much more interesting and enjoyable. Similar benefits can apply to shared research projects that are normally undertaken during the final year of a degree programme in the biological sciences (see Chapter 6).

3.8 Working on your own

As you progress though your degree programme, you should find that less emphasis is placed on laboratory classes in which you closely follow prescribed protocols for experiments. Instead you will be expected to take responsibility for the design and execution of your own experiments. This will apply, in particular, during final-year project studies. Final-year students often remark that they have not been asked to do anything quite like this before; they may feel exposed and unsure of how to proceed. In Chapter 6 of this book you will find detailed advice about how to make a success of your final-year project. In addition, the following strategy should help you build the confidence you will need as you begin to work independently in a laboratory. Essentially the strategy outlines how the Scientific Method (see page 42 and Chapter 6) can be enacted in the laboratory.

- **Plan and prioritise:** formulate null and alternative hypotheses and identify experiments that are likely to lead to meaningful results.

- **Protocol:** be clear about the procedures you need to follow and how they work. Think about the controls that will be needed, and how long the experiments are likely to take. Write your protocol clearly and concisely so that it will be easy to follow in the heat of the moment.

- **Prepare for the experiment:** set aside sufficient time, book the equipment, make up necessary reagents, re-read your protocol and make sure there are no mistakes.

- **Proceed with the experiment:** concentrate, be careful, take accurate readings, don't cut corners, make detailed notes in your lab book; if possible, record results directly in tables or graphs.

- **Process your results:** what do they mean? Are they consistent or inconsistent with the null hypothesis? Are they what you expected? If the data are not what you anticipated, try to think if anything might have gone wrong. However, it is very important to keep an open mind to the possibility of something new turning up – this is how discoveries are made! (See Chapter 1 on creativity in the biosciences.)

- **Produce a detailed record of the experiment:** include details of materials, the calculations used when making up solutions, the protocol, **all** the raw data (never discard any in case they turn out to be useful later) and your analysis. Finally, look back at previous experiments you have performed to see if you can identify any patterns emerging that could be included in the discussion of your new set of data.

Adhering to this scheme will provide a solid basis for you to develop an excellent scientific approach to your practical work. Do not, however, feel that you have to work in complete isolation. If you need advice when interpreting results don't hesitate to discuss your data with peers, demonstrators or tutors, and ask for help if you need it. Collaboration between scientists is essential: it enables the exchange and development of ideas. Furthermore, you may find that as you explain your results to others, new light is thrown on **your** understanding of the data. Of course, you should also be prepared to show an interest if others come to you for help and discussion of their results.

3.9 Writing-up experiments – the laboratory report

It is essential that you learn how to write a coherent report of laboratory work, which includes clear descriptions of experimental methods and results obtained, along with appropriate conclusions drawn from the data collected. Undergraduates often take some time to master these skills, but with practice you should rapidly become proficient in preparing laboratory reports, especially if you pay attention to advice and feedback provided by assessors (for more detailed advice on making the most of assessment and feedback see Chapter 10). The following information and advice, concerning the structure and content of a typical bioscience laboratory report, should help develop your writing and reporting skills.

3.9.1 Structure of the report

All laboratory reports need a clearly defined structure. They must start with an appropriate title followed by discrete Introduction, Methods, Results, Conclusion/Discussion and References sections (Table 3.2).

You may not be required to include a Methods section if the procedure you follow is clearly stated in your practical workbook. However, If you are required to deviate from the stated method this should be reported.

3.9.2 Writing style

Some students find it difficult to get the right balance between producing a report so brief that essential detail is lost and including pages and pages of description where the important points are buried in mountains of unnecessary information. At first, it is better to err on the side of providing too much detail (especially for your data) so that you do not miss any essentials; you can then adjust your style as you become more accustomed to what is required.

Table 3.2 Structure of a typical bioscience laboratory report

Section	Purpose	Suggested content
Title	To provide information on the content of the report.	One sentence that indicates concisely what the work was about.
Introduction	To set the scene. Why was the experiment done?	Relevant background information. Relate this to the aims of the experiment(s).
Method(s) (and materials)	To describe how the experiment was carried out.	Information about the materials used, procedures followed and any additional relevant information that alerts others to idiosyncrasies of the method followed.
Results	To present the data as clearly as possible.	A written description of the results with logically ordered tables, graphs, charts etc., to illustrate the data in the most appropriate manner. Calculations must be presented together with their workings. Include appropriate units.
Conclusion/ Discussion	To highlight the major findings of the experiment(s) and discuss their significance.	A discussion of all the results for accuracy, reliability and their corroboration, or otherwise, of current theories.
References	To indicate where information has been accessed.	Refer to textbooks, reviews and publications from journals.

3.9.3 Dealing with the results

A clear and concise presentation of results is central to a well-produced report. You will discover that there are many ways to present results, but the most effective involve tables, and figures such as graphs, charts or other diagrams. Sometimes it will be appropriate for you to include all your raw data (taken directly from an experiment) in your report. At other times mathematical analysis may be required to refine the data. Initially, you should try to get advice from your supervisor about the most appropriate format to use, but with practice you will soon know what is required. It is also worth noting how authors of articles in established peer-reviewed journals present their data, so that you can follow their example. This will help you develop good habits as you learn how to present data most effectively.

3.9.3.1 Points to note when preparing graphs

- Always include a descriptive title, and appropriate and clearly labelled axes (with relevant units).

- Avoid choosing awkward units of scale; for example, if you divide an axis into units representing 3 cm this will make it more difficult both to plot data and interpret the results.

- Use a sharp pencil to plot the data and different symbols for each dataset (with appropriate key) when plotting multiple sets of data on one graph.

- Do not extend a 'best-fit' line beyond the limit of the plotted data because you cannot make assumptions about data that have not been collected.

3.9.4 Tutor notes

Students frequently omit important details from their recorded observations; having noted the overall result, they may not realise the importance of detailing intermediary findings. They may be unable to give a full account of how one event led onto the next and therefore fail to appreciate the full significance of their results. This makes it difficult for them to provide a complete Discussion/Conclusions section. A related problem is that students make assumptions about what they think should happen rather than what has actually happened; so their reporting and analysis may be biased. These issues can be dealt with by giving timely advice at an early stage in the course.

Some students have great difficulty with graph plotting. The axes may be inappropriate and with unsuitable units or, more commonly, no units at all. Students occasionally become confused by units of time; for instance, plotting 1 min 30 s as 1.3 on an axis. By insisting that their first set of reports should contain only hand-drawn graphs, tutors will be able to check that all graphs contain the correct axes with an appropriate scale, suitable units and correctly placed data. Once all students have become competent at graph plotting, tutors might consider it appropriate to allow them to use computer software when presenting their data.

3.9.5 Discussion/Conclusions section

This section can cause real headaches for students unused to the rigours of scientific investigation. Having obtained results and presented them in an acceptable format, you must now draw appropriate conclusions. A crucial element of this is that you must be prepared to critically analyse your results and compare them to results and data previously presented by others.

Before you start writing this section, you should review the aims of the experiment and decide whether they have been fulfilled. Have all the questions been answered? If so, how have your results provided the answers? If the questions have not been answered, you need to be able to say why this was the case.

Next you must discuss the results you obtained (referring back to the appropriate figures or tables) and consider their significance. You should consider the accuracy of your results in terms of both human error (usually arising from pipetting and timing discrepancies) and any errors that may have been introduced by the equipment you used.

You may be able to compare your results with an expected range for the experiment, or with results from peer-reviewed publications. Discuss any reasons you can identify for differences between the results you obtained and the results of others (if appropriate provide references for published work that you cite – see below).

To finish, you should provide a summary of your conclusions and, if possible, any ideas you have on further avenues of investigation that could be followed.

3.9.6 Tutor notes

In the Discussion section, students sometimes place too much emphasis on causes of error (often, ironically detailing every possible factor that could influence their results but omitting a most obvious cause of error: poor pipetting technique) at the expense of a full discussion and interpretation of their results. They need to be encouraged to see the bigger picture from the beginning of their university career.

3.9.7   References

For any information you cite, you must provide details of where others can find that information. There are set conventions for citing references in the text and listing their details at the end of a report, paper, review article etc. The two styles used most commonly are the 'Harvard' and the 'Numbering' systems, but your university may expect you to use a different style and you should check details in your course guide, or with your tutor.

3.10   Concluding comments

Laboratory teaching is an essential component of biological sciences degree programmes that should educate students in the process of good scientific method. A key element of this is the development of the capacity to critically analyse data and fully understand the significance of results. Hopefully this chapter will help you build the confidence you will need to meet the challenges of practical work head on and ultimately enjoy the process of scientific discovery.

3.11   How you can achieve your potential in the laboratory

- Make sure you set aside time to prepare thoroughly prior to each practical class: this should involve background reading, and preparation of blank tables and graphs.

- Take a few minutes at the beginning of each class to become familiar with materials and apparatus.

- Ensure you fully understand how to use common items of laboratory equipment like automatic pipettes, precision balances and pH meters.

- Ensure you have a thorough grasp of the units of mass and volume used in the laboratory; in particular, you must understand the difference between units of 'M' and 'mol' (the first specifies a concentration, the second an amount).

- Understand and apply the Scientific Method during experimental investigations.

- Clearly record all results in a hardback laboratory notebook.

- Reflect critically on the results of your experiments; if your data do not conform to an anticipated result try to understand why: you might have made a mistake – on the other hand you may have discovered something really interesting.

3.12   Acknowledgements

We are very grateful to John Heritage for useful comments and suggestions, and thank Daniella Strauss for her input.

3.13   References

Adams, D.J. (2009) Current trends in laboratory class teaching in university bioscience programmes. *Bioscience Education* **13**, 3. www.bioscience.heacademy.ac.uk/journal/vol13/beej-13-3.aspx.

Archer, W. and Davison, J. (2008) *Graduate Employability: What do Employers Think and Want?* London: Council for Industry and Higher Education. Available at: www.cihe.co.uk/category/knowledge/publications/. Accessed April 2011.

Badge, J.L. and Badge, R.M. (2009) Online lab books for supervision of project students. *Bioscience Education* **14**, 1. www.bioscience.heacademy.ac.uk/journal/vol14/beej-14-c1.aspx.

Collis, M., Gibson, A., Hughes, I.E., Sayers, G. and Todd, M. (2008) The student view of 1st year laboratory work in the biosciences – score gamma? *Bioscience Education* **11**, 2. www.bioscience.heacademy.ac.uk/journal/vol11/beej-11-2.aspx.

Dawson, M.M. and Overfield, J.A. (2006) Plagiarism: do students know what it is? *Bioscience Education* **8**, 1. www.bioscience.heacademy.ac.uk/journal/vol8/beej-8-1.aspx.

Quinn, J.G., King, K. Roberts, D., Carey, L. and Mousley, A. (2009) Computer-based learning packages have a role, but care needs to be given as to when they are delivered. *Bioscience Education* **14**, 5. www.bioscience.heacademy.ac.uk/journal/vol14/beej-14-5.aspx.

Saunders, V. and Zuzel, K. (2010) Evaluating employability skills: employer and student perceptions. *Bioscience Education* **15**, 2. www.bioscience.heacademy.ac.uk/journal/vol15/beej-15-2.aspx.

Wilson, J., Adams, D.J., Arkle, S. *et al.* (2008) *1st Year Practicals: Their Role in Developing Future Bioscientists*. Leeds: UK Centre for Bioscience (www.bioscience.heacademy.ac.uk/ftp/reports/pracworkshopreport.pdf). Accessed April 2011.

3.14 Additional resources

3.14.1 General

The UK Centre for Bioscience provides a wide range of useful publications. These cover topics such as setting aims, and the design, organisation, supervision and assessment of laboratory practicals (www.bioscience.heacademy.ac.uk/ftp/resources/practical.pdf). Details of several practical sessions covering biochemistry, field biology, microbiology, pharmacology and biomedical sciences can be downloaded (and tailored appropriately according to the curriculum of the programme) from www.bioscience.heacademy.ac.uk/resources/themes/1styrpracticals.aspx. Accessed April 2011.

Two examples of useful texts that explain how to get the most out of practical teaching sessions:

Reed, R., Holmes, D., Weyers, J. and Jones, A. (2007) *Practical Skills in Biomolecular Sciences* (3rd Edition). Harlow: Pearson Education.

Jones, A., Reed, R., Wyers, J. and Martini, F.H. (2007) *Practical Skills in Biology* (4th Edition). Harlow: Pearson Education.

Millican, P. and Heritage, J. (2009) *Studying Science: A Guide to Undergraduate Success*. Bloxham: Scion. In a more condensed form, this describes approaches for students to adopt when tackling practical work and report writing as well as a host of other issues around university work.

The JISC Open Educational Resources programme provides a useful range of visual materials to which students can be directed to support their practical laboratory work. www.jorum.ac.uk. Accessed April 2011.

3.14.2 Specific practical techniques (accessed April 2011)

Using a Micropipette, GENIE CETL in the Department of Genetics and University of Leicester Audio Visual Services. Available at: http://www2.le.ac.uk/departments/genetics/vgec/educators/post18/topics/recombinanttechniques/micropipette-1?searchterm=using%20a%20pipette.

Introduction to Pipettes, created by Dr V. Rolfe, De Montfort University. Available at the JORUM website: http://open.jorum.ac.uk/xmlui/handle/123456789/2329?show=full.

Spectrophotometry, created by J. Koenig. Available at the JORUM website: http://open.jorum.ac.uk/xmlui/handle/123456789/372.

Using Spectrophotometry, created by J. Koenig. Available at the JORUM website: http://open.jorum.ac.uk/xmlui/handle/123456789/431.

Spectrophotometry, created by Dr V. Rolfe, De Montfort University. Available at the JORUM website: http://open. jorum.ac.uk/xmlui/handle/123456789/2330.

Introduction to Microscopy, created by Dr V. Rolfe, De Montfort University. Available at the JORUM website: http://open.jorum.ac.uk/xmlui/handle/123456789/2328.

The Light Microscope, created by Dr J. Heritage and Dr S. Bickerdike, University of Leeds. Available at the JORUM website: http://open.jorum.ac.uk/xmlui/handle/123456789/1339.

Basic Microbiology Techniques, created by Dr V. Rolfe and M. Ioannou, De Montfort University. Available at the JORUM website: http://open.jorum.ac.uk/xmlui/handle/123456789/2333.

Making and Running an Agarose Gel, GENIE CETL in the Department of Genetics and University of Leicester Audio Visual Service. Available at: www.youtube.com/watch?v=wXiiTW3pflM.

3.14.3   Calculations (accessed April 2011)

SI Units, created by R. Windle, School of Nursing and Academic Division of Midwifery, University of Nottingham. Available at the JORUM website: http://open.jorum.ac.uk/xmlui/handle/123456789/456.

Scientific Notation – Using Prefixes, created by J. Koenig. Available at the JORUM website: http://open.jorum.ac. uk/xmlui/handle/123456789/359.

Scientific Notation – Converting Between Units, created by J. Koenig. Available at the JORUM website: http:// open.jorum.ac.uk/xmlui/handle/123456789/360.

Equations – Calculations with Numbers and Symbols, created by J. Koenig. Available at the JORUM website: http://open.jorum.ac.uk/xmlui/handle/123456789/366.

Amount & Concentration – Making Solutions – Mole & Molar, created by J. Koenig. Available at the JORUM website: http://open.jorum.ac.uk/xmlui/handle/123456789/362.

Calculate Molar Absorbance Coefficient, created by J. Koenig. Available at the JORUM website: http://open. jorum.ac.uk/xmlui/handle/123456789/373.

Calculate the Concentration of a Substance, created by J. Koenig. Available at the JORUM website: http://open. jorum.ac.uk/xmlui/handle/123456789/334.

Dilutions, created by Dr J. Heritage and Dr S. Bickerdike, University of Leeds. Available at the JORUM website: http://open.jorum.ac.uk/xmlui/handle/123456789/1583.

Making Solutions: g/1% w/v, created by J. Koenig. Available at the JORUM website: http://open.jorum.ac.uk/ xmlui/handle/123456789/361.

Virtual Laboratories – Contents and Instructions, created by D. Male, The Open University. Available at the JORUM website: http://open.jorum.ac.uk/xmlui/handle/123456789/4479.

Biomathtutor covers basic mathematical calculations used in blood analysis and microbiology. Available at: www.bioscience.heacademy.ac.uk/network/numeracy.aspx.

3.15   Problems associated with Koch's postulates

- Perfectly healthy people may harbour potentially lethal pathogens. For example, *Streptococcus pneumoniae* is carried by many healthy individuals.
- Not all pathogens can be grown in pure culture. For example the bacteria that cause leprosy or syphilis cannot be grown in artificial media.
- Pathogens may have differing effects in humans and experimental animals. For example, *Salmonella enterica* var Typhi, the cause of potentially fatal typhoid fever in humans, causes mild diarrhoea when fed to mice, while its close relative *Salmonella enterica* var Typhimurium causes gastroenteritis in humans but a typhoid-like generalised infection when fed to mice.

4 Fieldwork

Julie Peacock, Julian R. Park and Alice L. Mauchline

4.1 Introduction

Fieldwork can be the most exciting and inspirational part of an undergraduate biology degree, but it can also be daunting, arduous and time consuming. In good weather it can be a pleasure, but in driving rain the experience is different and completing the fieldwork can become harder. However, by preparing correctly, all fieldwork can be ultimately rewarding.

The term fieldwork in this chapter will be used to cover work that takes place in an outdoor setting and covers everything from a short trip of just a couple of hours to a residential trip which may last a couple of weeks, or ongoing work which is undertaken over a number of visits.

Since fieldwork has the potential to be hard work and time consuming, it may be tempting, where possible, to avoid it in favour of something that delivers the same rewards in terms of marks towards your degree with the possibility of less effort. However the benefits and importance of fieldwork stretch far beyond the direct reward of degree marks, and provide an excellent opportunity to develop a wide range of subject-specific and generic skills.

Your fieldwork will link to other components of your degree course, and the subject-based knowledge you acquire during fieldwork will help you in these areas. You may even find it easier to remember and apply knowledge gained in the field rather than in a lecture or tutorial, as you are learning through hands-on experience. In addition, the data collection techniques and research methods you learn during fieldwork will help you when planning subsequent group or individual projects. When taking part in fieldwork you will be working closely alongside your peers and tutors; indeed on residential trips you'll be living alongside them too, and the interaction you have will be beneficial long after the fieldwork has finished.

If you aspire to undertake any form of field research or conservation work after your degree then it is essential that you make the most of every field opportunity available. It is also worth noting that a recent survey of fieldwork in bioscience undergraduate courses found that staff aim to provide opportunities during fieldwork for students to develop a wide range of skills, including the following: hypothesis development/testing; problem solving/organisation; identification; observation; use of field equipment/techniques; sampling/surveying; team work; application of statistics and data handling; report writing/presentation; safety (Maw *et al.*, 2011). Many of these skills are transferable and will therefore help with other aspects of your studies and future employability.

Effective Learning in the Life Sciences: How Students Can Achieve Their Full Potential, First Edition.
Edited by David J. Adams.
© 2011 John Wiley & Sons, Ltd. Published 2011 by John Wiley & Sons, Ltd.

4.1.1 Tutor notes

Running a field course is hard work and costly both in terms of money and time. However, with your guidance, students can benefit hugely, gaining both bioscience skills and skills from a 'hidden curriculum' that involves teamwork, interpersonal interactions, self-management and lifelong learning (Andrews *et al.*, 2003). The key to running a successful field course is to plan early and communicate clearly with students. Students may not have done this type of work before or visited the type of habitat you have in mind, so tell them everything they need to know. For someone undertaking fieldwork regularly, knowing what food, clothes, documentation etc. to take may seem common sense, but may not have been given careful consideration by students who will not know what to expect.

4.2 Fieldwork – exciting or overwhelming?

You may find the prospect of fieldwork exciting and something to be looked forward to. However, it is perfectly natural to feel nervous. This could be for many different reasons. You may feel unsure of how you will produce high quality work in an unfamiliar environment where you will also be required to learn new skills. Or, you may be concerned about working, and possibly living, with a group of students and staff you don't know well. It would be easy to let these concerns build up and create a mental barrier to getting the most out of the fieldwork. It is therefore important that you work out exactly what it is you're concerned about so you can address the problem.

If you are worried about working in an environment you've never been to before, find out exactly what kit to take with you, research the location and speak to your tutor. It will put your mind at rest to know you're well prepared. If you're concerned about the new skills you'll need to learn, remember you are likely to have done similar things before. If not, the tutor will probably start by teaching you what is expected, either at the start of the fieldwork or in sessions beforehand. He or she will provide handouts for you to read and point you in the direction of specific texts.

Perhaps you are a technophobe who is worried about using some unfamiliar hi-tech equipment in potentially challenging environments. Pay attention when you are shown how to use the equipment and read any notes given. If you are working in groups it is likely another group member will be able to help you with unfamiliar equipment. If they can't or you are working individually, ask a tutor or demonstrator to help. Alternatively, try borrowing the equipment before you do the fieldwork, take it outside and practise. Laboratory technicians are usually happy to ensure project students know what they are doing, especially with expensive kit!

It could be the social side of fieldwork that you find daunting; perhaps you don't know anyone else in the group. If so, make an effort, during pre-fieldwork meetings, to get to know one or two of the others or start a social networking website associated with the event. Once in the field you will quickly get to know people as you work together. If communal cooking is a worry because of food allergies or for religious reasons, speak to the fieldwork organiser as soon as possible to ensure your needs are met. Similarly, if dormitory or other sleeping or washing arrangements are of concern to you, then speak to the organiser. They should be able to put your mind at rest and come up with a suitable plan. You could also look up the website for the planned accommodation or phone in advance to find out further information that will help you feel more comfortable about the arrangements.

Finally, remember that whatever your concerns, however confident other students may appear, they are likely to be wondering about the same things! By raising your concerns you are likely to help put their minds at rest.

4.2.1 Tutor notes

A recent study by Boyle *et al.* (2007) showed that many students were apprehensive prior to fieldwork. Briefing students (about accommodation arrangements etc.) as early as possible can help allay anxiety. Remember, there may be a need to accommodate various cultural and religious needs when on fieldtrips. For example, prayer space and somewhere to wash for Muslim students; separate sleeping areas for men and women; and not basing all social activity in the pub. Developing social bonds between students is an important outcome from a fieldtrip, but this may not happen automatically (Nairn *et al.*, 2000). Therefore, 'icebreakers' may be needed at the start of the course. For example, a 'speed dating' icebreaker could be used, with students in an inner and an outer circle. Students in the inner circle are given one minute to discuss with the nearest student in the outer circle what they enjoy about fieldwork. The inner circle moves one place to the left and the exercise is repeated as often as time permits. Alternatively, you could organise a field-based exercise where small groups have to go quickly into the field and identify, and if appropriate retrieve, objects. For example leaves of certain trees, or they might take photographs of certain species.

4.3 Planning and time management

Practical laboratory experimentation requires planning and time management to ensure that the work is conducted using a logical scientific approach over a realistic time frame. Fieldwork demands a high level of planning as it is often conducted at a distance from your institution, and it is not time-efficient to travel backwards and forwards to collect forgotten pieces of equipment. Fieldwork is also often conducted in unfamiliar environments and consequently problems occur, many of which are due to the outdoor nature of the work. Try to envisage potential problems in advance and work out ways around them so that you are prepared for what would otherwise be the unexpected. This section describes some of the most important considerations for planning field experiments and is divided into three sub-categories: before, during and after fieldwork.

4.3.1 Before fieldwork

Plan ahead! If you are going on an organised trip then you need to be clear where you are going and what you will be doing when you get there. If you are planning your own project, then you need to know exactly what data you need to collect to answer your research questions, and you will need to prepare a detailed work schedule before setting off for the field. It will help to speak to others who have completed similar field tasks as they will be familiar with the time the work takes and any specific issues you should be aware of. If you are working in a group you need to establish individual responsibilities and the role you are expected to play as part of the team in achieving any collective goals. It is also important to bear in mind any specific assessments related to the fieldwork (see Section 4.12, Feedback and assessment).

You need to plan how you are going to get to the field site and allow sufficient time for travel. Work out the route in advance and, if arranging to meet others, plan a time and place but also have a contingency plan ready in case of problems. You should avoid working alone.

Your planning stage should include finalising the experimental approach, collecting and assembling equipment, preparing record sheets, labelling storage materials, and completing risk assessments. All field equipment should be checked to ensure it is working and the batteries are charged. Clothing should be chosen to make the fieldwork as comfortable as possible. Minor things can also make the work much more enjoyable; ensure: your mobile phone is charged, you have plenty of food and drink and you have any medication required.

Below is a general, alphabetical list of kit for fieldwork:

All handouts provided in advance	If you laminate these they won't spoil if raining and you can wipe off the dirt.
Camera	Plus spare battery and films/memory cards.
Clipboard	It may be worth taking one that is weather proof – i.e. it should have a cover that will allow you to write in rain or snow without getting the paper wet.
Field guides	Plans and biological keys as necessary.
First-aid kit	Appropriate to the fieldwork you are doing. Know how to use the contents.
Food and drink	Plus emergency rations – should circumstance mean you have to stay out a lot longer than expected.
Maps	And compass and/or GPS.
Mobile phone/radio	With any contact numbers you need (emergency, people you are meeting, University contacts) programmed into the phone and written down separately; make sure the phone is fully charged. If you are doing fieldwork for a number of days, make sure you have a way of charging it. Use a two-way radio if no signal is available for a phone.
Notebook, any pre-printed datasheets and pencils	In wet weather a pencil will keep working ... a pen won't. If you're going somewhere particularly humid or wet, consider taking waterproof paper.
Protective clothing and equipment	Waterproof clothing, including footwear, is very important to remain comfortable. Even if not raining, dense undergrowth can remain damp. Also pack sun cream and sun hat for hot days.
Rucksack	For carrying the kit.
Specialised equipment for the task	Check you know how to use it and that it is working.
Storage materials	For example, re-sealable bags, marker pens and labels. If possible label storage materials before you go on the trip.
Torch	With spare batteries.
Whistle	Remember the international distress signal is six long whistle blasts, torch flashes, arm waving or shouts for help, in succession, followed by a one-minute pause before repeating. The response to a distress signal is three short whistle blasts, followed by a pause before repeating.
Anything else you were told to take	

Figure 4.1 Group briefing and site familiarisation prior to survey work

4.3.2 During fieldwork

With careful pre-planning, your fieldwork should run smoothly. Keep a check on time as tasks often take much longer than expected and you may need to adjust your schedule. It is worth having contingency plans in place in case the original strategy doesn't work. If you do have to change your plans in the field, ensure that you are still going to be able to complete the task set or answer the research question posed. If you feel this will not now be possible, you should try to speak to your tutor before making the changes. Additionally, if you are changing field site or spending longer in the field than planned then you must tell all the relevant people (see Section 4.8, Safety and permissions). If you are working as part of a group, make sure you communicate with the other team members and ensure that everyone is aware of any changes to the overall plan (Figure 4.1).

4.3.3 After fieldwork

Although the largest chunk of time probably needs to be allocated to conducting the fieldwork, this shouldn't be at the expense of having time available at the end of the day to collect your thoughts, file datasheets, write up field notebooks and, if necessary, finalise plans for the following day. This is best done when the day's work is still fresh in your mind, as, although you might think you will remember what happened, it is very easy to get muddled unless accurate records are made at the time of data collection. It is also important to ensure that all samples are adequately labelled and stored. Biological specimens can degrade quickly and therefore you should consider appropriate temporary storage methods (such as refrigeration) if there is insufficient time to process all your samples immediately.

If the fieldwork is to continue the following day, then many of the tasks listed in the 'before fieldwork' section need to be repeated; the equipment needs to be washed and reassembled, new record sheets prepared, vehicles refuelled, new route planned etc.

The importance of accurate labelling and recording cannot be over-emphasised. Many field sessions are compromised because accurate post-collection procedures were not completed correctly.

4.3.4   *Tutor notes* ───

Recommendations for good practice in fieldwork can be found in Gold *et al.* (1991) and Kent *et al.* (1997). The fieldwork should be placed in the context of previously taught course elements, as this will help students make learning connections. Pre-course preparation material can help reduce the 'novelty effect' of fieldwork by helping students become familiar with a new environment (Cotton and Cotton, 2009). Often time can be a limiting factor during fieldwork and there can be a tendency to devote most of the students' time to data collection. It is important to allocate sufficient time to the planning phase and to allow students time to develop their research ideas and form hypotheses. Providing clear learning outcomes for each fieldwork task is important to help students plan effectively (see also Chapter 10). Finally, students should be asked to submit a plan of their field protocol well in advance of the fieldwork: this will promote good practice by encouraging them to plan in advance and will help ensure that they are well prepared.

4.4   Group work and social aspects of fieldwork

During lectures it can be difficult to get to know other students. Attending a fieldtrip is therefore a good opportunity to get to know your peers better and meet new people. Such shared experiences can often be the start of lasting friendships. The informal atmosphere on fieldtrips also allows you to get to know the teaching staff better. Establishing a closer relationship is beneficial as you will learn about their research interests and may find it easier to approach them for help later in your course.

Working together on shared tasks in the field is a good way to develop working relationships with others. The experience of being in an unfamiliar environment and having to work with others to achieve a shared goal requires multiple social and other skills. These include: listening, problem solving, team working, leadership, methodological thinking, communication, presentation, planning, recording, observation, project management, time management and the ability to reach a compromise. The Learn Higher website provides some excellent videos to support group work (see Section 4.16, Additional resources). As a group member it is important to work and communicate with other members from the start of the process. It is useful to appoint a team leader (on a 'rotating' basis). When it is your turn you should consider the following:

- Ensure each group member is introduced to the other members of the team so you all know who is in the group.

- Identify group members with specific skills and, conversely, those who may need additional support learning a skill or technique.

- Confirm everyone is clear about the task objectives and sampling protocols.

- Check each group member knows the part of the task for which he or she will have responsibility.

- Ensure everyone understands how long the group will have to complete the task, and make sure the timeframe is realistic.

- Make sure everyone stays focused on the task.

- If it is an ongoing project, decide how group members will contact one another.

It is important to develop the ability to think ideas through and communicate them to others. Good communication is essential in any team. Often the group will need to find a way to reach a compromise between several contrasting opinions, and clear communication of ideas will be important during this process. Acquiring such skills, while learning about your subject, will be hugely beneficial and will stand you in good stead when faced with similar situations in the future.

4.5 Collecting the right data

What are the 'right' data? This entirely depends on the question you are asking or the biological process you are monitoring. The following structure for experimental design can help you plan your approach. In essence it involves application of the Scientific Method (see Chapter 6, Section 6.3) in the field.

Research process:

(1) Decide on the research question and develop hypothesis.

(2) Refer to scientific literature.

(3) Decide how to test the hypothesis.

(4) Devise a field protocol and sampling plan.

(5) Think about treatments, controls, blocking, replication, randomisation and time availability.

(6) Consider how to analyse the data.

(7) Prepare an equipment list and fieldwork schedule.

(See the worked example of the development of a field sampling strategy, Section 4.5.1.)

Various textbooks and study guides are available that provide useful starting points when designing fieldwork (see Section 4.16, Additional resources, for suggestions). Take some time to become familiar with the methods described in these texts.

Important questions that arise as you consider and 'unpick' your research question relate to the field sampling strategy, the design of your experiment or monitoring exercise, and issues associated with replication. You should consider the type of statistical analysis you wish to use to interpret your data **before** starting the fieldwork; this will help inform you of the amount and type of replication you need to be able to draw appropriate conclusions. Analysing field data can be demanding because it is difficult to control for variables in the same way you would during a laboratory project. Dytham (2003) provides a checklist to help you decide which statistical tests to use, and describes how to perform these tests using several common statistical software packages (see also Chapter 7 and the *Engage in Research* website under Additional resources). Discuss data analysis with your tutor, or other statistical expert, at the planning stage, as your conclusions will be limited if you've collected inadequate data in the field.

When in the field, careful data collection (to the appropriate level of accuracy) and recording are essential if you are to reliably test your hypothesis. In Section 4.6 we describe the use of technology that can enhance and often speed up data collection in the field. However, whichever method you use, you must have a robust recording scheme to prevent mistakes creeping into your data.

4.5.1 Worked example 1 Development of a field sampling strategy

Using the research process outlined above, an investigation of the distribution of lichen on a rocky shore (Figure 4.2) might be described as follows:

(1) How does lichen distribution vary on a rocky shore?

Null hypothesis = lichen species distribution does not vary with distance from the shoreline.

Alternative hypothesis = lichen species distribution does vary with distance from the shoreline.

(2) Gather lichen identification guides.

Look up previous studies asking similar questions (not necessarily about lichens).

(3) Record abundance of lichen species at fixed distances away from the shoreline on a rocky beach.

(4) Identify accessible rocky beaches and mark three line transects up from the shoreline on each beach (make these as representative of that site as possible).

Decide a sampling strategy; i.e. record all lichen species present within a $1\,m^2$ quadrat every $10\,m$ up the transects.

(5) Treatment = distance from shoreline

Controls = n/a

Block = transect

Replication = beach (as many as possible in the time available; avoid pseudoreplication)

Randomisation = throw the quadrat randomly near to the sampling point (to avoid bias in choosing a 'nice looking' area to sample)

Figure 4.2   Student group using a quadrat to investigate the distribution of organisms on a rocky shore

Time availability = collect as much data as possible, but if beach is very large then adjust sampling to every 20 m or sample fewer beaches (but be mindful of the need for replication).

(6) ANOVA on abundance data.

Diversity indices.

(7) Maps, clipboard and pen, lichen identification guides, digital camera (to photograph species for later identification), tape measure, quadrat, basic fieldwork kit (see generic list, Section 4.3.1).

For further explanation of the statistical terms used and how to analyse fieldwork data see Dytham (2003), the sections on Planning Your Research and Step-by-Step Statistics at the *Engage in Research* website www.engageinresearch.ac.uk/, and Chapter 7 of this book.

4.5.2 Tutor notes

A lot of students' anxiety about the academic side of fieldwork is associated with the use of unfamiliar mathematical and statistical concepts (Cotton and Cotton, 2009). Field data may require more complex analysis than students have previously been taught. Pre-course preparation can help encourage students consider how they will analyse their data before they start the fieldwork. Ensure there is sufficient time for data analysis when students return from the field. See Chapter 7 for more ideas.

4.6 Technology in the field

This is an expanding area, as new technology is continually being developed and made more affordable and robust for use in the field. Information technology (IT) is increasingly used during fieldwork courses for various reasons, but primarily because it increases efficiency of data collection and collation and is an important part of the skills that you develop while out in the field (Maskall *et al.*, 2007). Commonly used IT tools include: personal digital assistants, tablet/notebook computers, global positioning system (GPS), digital cameras, video cameras, mobile phones, data loggers and laser scanners.

Mobile phones are now an important part of fieldwork kit. They can be used to facilitate good team communication even if team members are widely dispersed. They provide a way to instantly contact emergency services if needed, and allow coordinators to check up on the safe working of individuals while out on fieldwork. Mobiles with advanced functions such as internet access and GPS tracking can also be used to access a large amount of information from remote locations.

Digital cameras, and sound and video recorders, can produce useful reference materials and can record evidence for later investigation (for example, photos of unknown insect species can be identified in the lab using taxonomic keys etc.). Automatic data loggers are small, battery-powered, electronic devices that record data over time using internal or external sensors and instruments. They can be left unattended in the field to monitor temperature, humidity and other environmental parameters; the data can then be downloaded to a PC using a wireless or USB connection. Laser scanners are an alternative to traditional survey techniques and can be used to collect surface data to provide 3D spatial orientation.

Figure 4.3 Student using a clinometer in poor field conditions

Another area where technology can greatly assist in fieldwork is in spatial mapping involving the use of GIS (geographic information system) data. Hand-held GPS units are available that can accurately pinpoint your position anywhere in the world, allowing you to record sample points very precisely and with spatial reference to one another. GIS applications can provide a way to combine layers of biological data with reference to geographic location. This enables analysis and mapping of spatial data.

In addition to the generic technological equipment already described, there is a wide range of specialist technology/equipment available depending on the subject under investigation (Figure 4.3). Before using this equipment in the field, you need to check the devices and calibrate them if necessary. You also need to ensure that the correct version of any software is uploaded and that you are familiar with how it all works (otherwise valuable field time can easily be lost in setting up new equipment).

Away from the field, Web 2.0 technologies provide ways to share information. If you are working as a member of a group you could develop a wiki that will allow you to share data files, discuss research findings and coordinate the project. This could be an exciting way to work towards the final report for the project and will provide a way for others to browse (and assess) your output. A blog could be used to share information and as a way of recording your reflections of time spent in the field.

4.6.1 Tutor notes

The use of technology has many useful implications for teaching in addition to its use in the practical side of fieldwork. Computer-based assessments during a field-based module have been shown to enhance self-regulated learning and provide an opportunity for students to widen their knowledge base (Baggott and Rayne, 2007). Personal digital assistants (PDAs) can also be used in the field to allow data to be collated directly, thereby saving time and allowing students to focus on the main learning outcomes (Baggott, 2009). Remember to check well in advance that: there are sufficient PDAs for your group size; you know how to use them; and you have a back-up plan if they don't work. Setting up a class wiki will enable students to compile group data and share ideas, and will allow you to monitor progress of project work.

4.7 Costs, sustainability and ethics

Fieldwork is expensive in terms of time. It can also be costly in financial terms due to the need for tuition, accommodation, subsistence, insurance, specialised field equipment, materials etc. Fieldwork can be especially expensive if travel or access to remote areas is required. You may be asked to pay some or all of these costs, so be sure to enquire about all the related expenses and whether you would be eligible for any grants before signing up for a field course. Many Universities will have funds available to help cover the costs of compulsory fieldtrips for students who are in financial hardship.

It is important that you conduct fieldwork in an ethical and sustainable manner. When planning your work, consider how you will do this. For example, overseas fieldwork is not only expensive but produces a large carbon footprint as well. Local fieldtrips may be just as advantageous in terms of learning outcomes, but more sustainable in terms of financial cost, time and environmental concerns. Some people may have ethical objections to the fieldwork itself if it involves destroying part of a habitat or collecting live specimens. Try to develop a research project with minimal environmental impact; for example, ensure that any biological monitoring programme you are planning does not result in damage to the population or resource you are recording.

4.7.1 Tutor notes

Fieldwork can be more costly than other methods of teaching; however, Maskall and Stokes (2008) provide a useful discussion on ways to reduce its financial costs. Overseas field courses can provide a dramatically different learning environment and be a great perk for students and tutors alike. However, they incur greater financial, time and environmental costs compared to local fieldtrips. Consider whether the learning outcomes of the overseas trip justify the extra cost.

The UK Centre for Bioscience has produced a useful 'How to' sheet on making fieldtrips more sustainable (www.bioscience.heacademy.ac.uk/ftp/esd/howtofieldtrips.pdf). It is important to involve students in efforts to increase the sustainability of fieldwork, as they are then more aware of its environmental impact. Mark Huxham discussed the ethical issues involved with partici-pating in a field course in the Algarve with his students (www.bioscience.heacademy.ac.uk/ftp/events/cardiff091209/huxham.pdf), and one of the learning outcomes was to 'evaluate the ethical issues involved in overseas fieldwork, decide whether carbon offsetting is appropriate and if so make a commitment to a particular method'. The students chose to participate in the field course, but also contributed to tree planting near their university.

4.8 Safety and permissions

4.8.1 Safety

it is better to be a live donkey than a dead lion

Sir Ernest Shackleton's comment to his wife on
having turned back only 97 miles from the South Pole in 1909

The most important aspect of any fieldwork is the health and safety of all those taking part. Although the tutor in charge of your fieldwork will have overall responsibility for this, it is important to remember that all participants have a duty of care to themselves and others. If you are undertaking an individual research project that involves fieldwork you will have to plan carefully to minimise all risks and fill in a risk assessment form. Most risks can be reduced through common sense and careful planning. Your institution will have its own rules which you must follow at all times. Some general points:

- Follow all written and verbal advice; i.e. where to go/not go, what equipment to bring, etc.

- Make staff and others in your group aware of your medical circumstances, for example whether you suffer from asthma, epilepsy or diabetes, and carry medication and medical details as necessary.

- Check the weather forecast and consider postponing or be prepared for postponement if conditions are extreme.

- Your clothing and equipment must be suitable for all the weather conditions likely to be encountered. If you are working on the hills, ensure you are prepared for cold, wet conditions even if it seems calm at lower levels. If the forecast is for sunshine, take a hat and sun cream. Wellies may be perfect for boggy conditions but may be slippery on wet rocks, where sturdy boots may be preferable. Take advice from those who have worked at the site before.

- Carry a comprehensive first-aid kit and know how to use it.

- Depending on your location it may be important to know how to read a map and use a compass; make sure you have them with you.

- Leave details of your intended work locations, routes, expected times of departure and arrival, and emergency contacts; ensure the person who holds this information knows what to do if you don't return at the expected time.

- Report safety-related incidents immediately to your course leader; examples could be getting lost finding, or returning from, the field site, or minor injuries.

- Remember the international distress signal on land is: six long whistle blasts, torch flashes, arm waving or shouts for help in succession followed by a one-minute pause before repeating. The response to a distress signal is three short whistle blasts, followed by a pause before repeating.

• Check your work is covered by your college or university insurance, and make suitable arrangements if it is not.

The following case studies (Sections 4.8.2 and 4.8.3) are based on projects that students at UK universities have carried out. Think about them carefully; how might **you** reduce the risks involved?

4.8.2 Case study

A student plans to carry out fieldwork at a woodland nature reserve that is a five-minute walk from a main road near a city centre. The task will be to record tree species present, measure tree diameter and calculate tree height. There will be no need to climb the trees to make the measurements. The student plans to carry out the work on one day, between 9 a.m. and 4.30 p.m., in February, and is able to complete the work without assistance, so will therefore work alone. Risks and how they may be reduced can be considered as shown in Table 4.1.

You can see that similar solutions can reduce a range of different risks; so making small changes to the plan can greatly improve the safety of those involved with the fieldwork. Can you think of any additional risks associated with this project?

Table 4.1 Risks and how they can be reduced

What are the risks?	Why a risk?	How can the risks be reduced without affecting the study?
Working alone	This is a risk even at a site relatively close to a main road. • Members of the public pose a risk should they wish to cause you harm. • If you have an accident there is no one to help.	• Ideally you should work in a group of three or more. You may be able to arrange your fieldwork so that you are accompanied by peers who are doing their own research at the same location, or ask friends to come along. • Carry a mobile phone (check it will work at the site) or radio so you can contact others. • Work in daylight hours; ideally the field site should not be in a secluded area. • Do not tell strangers you will be working there all day by yourself. • Carry a personal alarm, whistle and torch to summon help.
By 4.30 p.m. in February it will be getting dark	Working in darkness, or even dull light, will be harder. • You are more likely to have an accident such as tripping over an obstacle.	Ideally you should arrange your work so that you will be back before dark. • This could be done by dividing the work over two half days – working over two successive days is unlikely to affect the results obtained for this study.

(continued)

Table 4.1 (*Continued*)

What are the risks?	Why a risk?	How can the risks be reduced without affecting the study?
	• You are also more at risk from harm from members of the public and there are likely to be fewer people around to help you. • In February it is likely to be cold and the temperature will drop further at nightfall – see below.	• Working at a different time of year would enable a longer working day in light conditions. In this study the results are unlikely to be drastically altered by working in late spring or summer, and the tree species will be more easily identifiable with leaves. • If the above options aren't possible, taking someone to assist you with recording data may speed up your work and enable you to get home in daylight.
Cold temperatures	When walking or doing physical work outside in February you may feel warm; however when standing around taking and recording measurements you are likely to get cold quickly. • This will make it harder and more unpleasant to work and you're more likely to have an accident such as tripping. • Getting extremely cold could lead to hypothermia.	• Make sure you have plenty of warm and waterproof clothing with you; layers are better than one thick jumper. • Carry a flask with a hot drink in it. • In your first-aid kit you should carry a survival blanket: if you are injured it will keep you warm. • You could consider doing the fieldwork at a different time of year or organising the work in short sessions so you don't get cold.
Tripping in woodland	It would not be unusual for this to happen and the injury from such an accident is likely to be minor. However: • If you twist your ankle then walking even the short distance back to the main road may be difficult. • It is also possible you could trip into a stream or sustain a head injury.	• Pay attention to where you are walking. • Work in good light conditions so you can see trip hazards. • Wear walking boots which will support ankles in the event of a fall. • Carry a first-aid kit and know how to use it to treat injuries.

Table 4.1 (*Continued*)

What are the risks?	Why a risk?	How can the risks be reduced without affecting the study?
		• Have a mobile phone/radio with you to get help if needed. • Work in a group of three or more, so one group member can go for help while the other(s) wait(s) with the injured party.
Tree/branch falling on student	This is unlikely to happen, but if it did could cause serious injury.	• The chances of such an accident would be increased in a storm and it may be worth postponing fieldwork in these conditions. • Carry a first-aid kit, mobile phone or radio and ideally work in a group.
Risk posed by domestic or wild animal	Domesticated animals may not be permitted in a nature reserve, but you may encounter dogs or cattle when walking to and from the reserve. Wild animals in the UK that are most likely to cause harm, such as bees, wasps and snakes, are unlikely to be encountered in February.	• If you have a known allergic response to bee stings then you should carry your prescribed EpiPen. • Get help immediately in the event of a snakebite.
Risks posed by litter	The nature reserve is close to a main road near a city centre. Litter may have been dropped in the area and this may include contaminated needles.	• Check the area carefully. • Wear padded, protective gloves. • Seek medical advice if you are injured by a needle.

4.8.3 Case study

Work through the following scenario:

A student is planning to carry out fieldwork to study crustacean populations along transects on three rocky shores in North Wales. The shores are only exposed at low tide. The student plans to take a friend to help set up the transects and take notes. They plan to carry out the fieldwork over a three-week period in the autumn.

Detail the risks involved and suggest ways to reduce these. Can the original plans be altered to reduce risk without affecting the quality of the science?

4.8.4 Tutor notes

Getting students to take health, safety and risk assessment seriously can be difficult, as they tend to find these topics dry and boring. Ask students to fill out risk assessment forms for all of their fieldwork even though, for group trips, you will have completed the official form. Once students are accustomed to completing these forms they will be more aware of the risks associated with fieldwork and will be more competent when completing risk assessments for their own projects in the future.

You could ask students to submit a paper or electronic risk assessment before the fieldtrip and make it a requirement for completing the course. You would not necessarily need to mark these individually, but once submitted you could provide students with the actual risk assessment you've completed to compare with their own. Alternatively, you could set up a class or group wiki for students to fill in a risk assessment communally, with a requirement that all students have to contribute something. Once the students have finished you could add comments. You never know, the students may even come up with something you've never considered. By showing that you take health and safety seriously, the students will as well.

4.8.5 Permissions

When planning fieldwork it is important to get permission from the landowner, since not doing so may lead to you trespassing on private property. Explain to the landowner who you are, where you are from, exactly what you plan to do, and when. Many landowners will be interested in your work and may be able to offer helpful advice.

When contacting landowners you should ask about:

- livestock, or ground nesting birds you should avoid;

- timing of any shooting or crop spraying;

- location of slurry pits (including disused ones) or old mine shafts; and

- any hygiene precautions they want you to follow and whether they have facilities you could use to meet their requirements (for example, areas where you could wash down vehicles or clean boots).

Access rights and restrictions on certain activities are set out in the Countryside and Rights of Way Act (2000) for England and Wales and in the Scottish Outdoor Access Code (see Section 4.16, Additional resources). Protected areas such as Sites of Special Scientific Interest (SSSIs) may have further restrictions on access and the types of biological specimens you are able to collect.

Certain locations are extremely high-risk environments; you will not only need to think carefully as to whether the scientific research is worth the safety risk but you will also need to gain formal permission from the relevant authorities. Such locations include:

- airfields;

- construction sites;

- landfill sites;

- Ministry of Defence property;

- areas bordering motorways; and

- quarries.

The *NERC Guidance Note: A Safe System of Fieldwork 2007* provides more detailed safety information (see Section 4.16, Additional resources).

4.9 Accessibility

You may have a condition that you feel will prevent you from doing fieldwork, or that will make it harder for you to work in the field. It is important that, whatever your concern and however trivial you feel it may be, you should speak to the fieldwork organiser as soon as possible. He or she will work hard to take all enthusiastic students with them on a trip, but needs to be prepared. It is also important to discuss with your fieldwork organiser, or disclose on health and safety forms, health issues you would normally deal with by yourself and not consider a problem. In the unlikely event of you becoming unwell while undertaking fieldwork, the staff will then be well prepared.

You must remember that while the course organiser is an expert in fieldwork, you are the expert when it comes to knowing what it will take to enable you to do the work. Communication is the key to making sure that everyone is happy and you are able to achieve the proposed learning outcomes. Don't worry about being a nuisance; it is likely that a solution to your problem will benefit others in the group, or even the whole group. For instance, if you are dyslexic, you may need to ensure that all handouts and any forms you need to fill in are obtained well in advance of the fieldtrip. Getting information in advance will benefit everyone. Alternatively, for a student who requires daily dialysis, going on a residential fieldtrip to a remote location may not be possible. Instead, a series of local trips may be better and this option may also suit students who have family or work commitments, or a tight budget (Healey *et al.,* 2001). So, by speaking to the organiser early on you may improve the fieldwork experience for the whole group.

When talking with your course organiser, find out exactly what is planned and how it relates to the learning outcomes. For example, is the walk of one mile to the field site an essential part of the fieldwork or just a way of getting there? Perhaps a coach cannot get any closer, but a car might. This alternative may benefit you if you have reduced mobility, but could also aid those who have severe asthma. Similarly, the route taken may have been chosen because it is the most scenic, but is over uneven ground with several stiles. An alternative route, over even ground with gates rather than stiles, may be far more accessible for you. A quick conversation before the event could prompt the fieldwork organiser to make simple changes that ensure you get the most out of the fieldwork.

If you have, for example, epilepsy, asthma or diabetes, you may feel you can cope with the condition on your own, as you usually would. However, it is important that you speak to your fieldwork organiser to make sure all arrangements are in place. For example, if you are insulin dependent and you need to store your insulin in a fridge, you must not assume that a fridge will be available. Check first; arrangements may also need to be made for the disposal of sharps. In addition, the fieldwork organiser may not have considered it necessary to plan for regular food/meal breaks. Again, speaking to the organiser beforehand will allow for the inclusion of more, or longer, meal breaks in the schedule. Asking the course organiser to provide a detailed itinerary will not only help you but also those students who get anxious when facing unknown situations. If you suffer from conditions such as diabetes, epilepsy, asthma or depression, it would be a good idea to make the course organiser aware of any known triggers or allergens (e.g. pollen, tiredness, strenuous activity) and plan how these could be avoided or reduced. Additionally you may want to ensure you are not going to be working on your own in a remote location, but in a group of at least three students with communication links to the other students and staff, or external help.

If you usually have help from a support worker, and think you will need one with you during fieldwork, you will need to let the course organiser know well in advance. Travel, accommodation and insurance for your helper will all need to be organised. Similarly if you have a guide dog, don't assume the fieldwork organiser will take this into consideration without you mentioning it; you will need to ensure the dog's needs can be met. Let the fieldwork organiser know that the costs of special arrangements such as a support worker, or special equipment, may be covered by your Disabled Student's Allowance.

You may wish to suggest, when speaking to your fieldwork organiser, that they read the most appropriate of the six web-based guides in the series *Learning Support for Disabled Students Undertaking Fieldwork and Related Activities* published by the Geography Discipline Network (see Section 4.16, Additional resources). Although designed for geographers, they also apply to fieldwork in the biosciences. The guide(s) will also give you a better idea of what might be done to make fieldwork more accessible for you.

Making sure that you are not at a disadvantage during fieldwork is a process of negotiation. Speak to the course organiser as soon as possible to ensure you are able to meet the learning outcomes. However, it should also be noted that while there may be elements of fieldwork that you and other students would rather not do, personal issues should not be used as invalid excuses for avoiding perceived unpleasant or demanding tasks.

4.9.1 Tutor notes

When planning a fieldtrip it is important to think from the outset about accessibility issues. Simple steps such as providing detailed information about the trip as early as possible, both online and as hard copy, will enable students to work out if there is anything that may disadvantage them when taking part. In 2007/2008, 7% of undergraduate students declared a disability (source: the Higher Education Statistics Agency, www.hesa.ac.uk/, 2009); as students are not obliged to do this, the true number is likely to be higher. The law states that they must not be disadvantaged (Disability Discrimination Act 1995). Of students who declared a disability, 44% identified a specific learning issue (such as dyslexia, or dyspraxia). This means that over 3% of all students are affected by these conditions. When preparing handouts you can help these students by using a sans serif font (e.g. Arial) of a minimum 12-point font size. For more detailed information about their requirements see Chalkley and Waterfield (2001), look at the JISC Techdis website (www.techdis.ac.uk) or contact your institution's Disabilities Office. Fifteen percent of students who declared a disability declared an unseen disability, which would include epilepsy, diabetes and asthma. You may not normally need to be aware of these illnesses, but during fieldwork they could become an issue. It is worth being aware of the basic first aid that could be needed and of other arrangements you may need to make. Six percent of students who declared a disability declared mental health issues. Mental illness is a disability students may be less likely to declare, and the true number of students with such issues could therefore be much higher. Fieldwork in unfamiliar surroundings, which is physically and mentally challenging and removed from a student's usual support network, may exacerbate mental health issues. Of course, students coping with mental health issues may have no problems during fieldwork. However, if they do, it may be hard to predict the best way to deal with the situation. It would be much better to be prepared and be able to support these students from the outset; to do that you need to encourage the students to speak to you (Birnie and Grant, 2001).

You cannot make a student disclose a disability, but if he or she thinks you can help they are more likely to come forward and you will then be able to plan the trip so it works for all students. Give examples of the sorts of conditions they should tell you about, such as epilepsy, asthma, depression. If you explicitly mention a mental health issue, students may be more comfortable about disclosing their own condition (Birnie and Grant, 2001). State that the information given will not be passed on to anyone else without the student's permission. Give students a time when they can see you in your office; they may be uncomfortable about discussing health issues at the end of a lecture when other students are around. Get students to return health and safety forms early enough for you to act on any issues raised. Give a deadline after which you can't guarantee to have appropriate support available, so you don't have to make arrangements at the last minute. Once support is in place, ensure students are fully aware of the provision you have made so they don't worry unnecessarily.

There are many other ways you can think about making fieldwork accessible at the planning stage. Read the detailed advice provided by Healey *et al.* (2001), speak to your university's Disabilities Officer, consider attending a course on accessible teaching and, before the course starts, speak to the students themselves so that together you can devise fieldwork suitable for all involved.

4.10 Making the most of different types of fieldwork

4.10.1 Short excursions

All fieldwork is time limited, but when the trip is for just a few hours it is particularly important to use the time in the best way possible. Short trips will often take the form of a visit to somewhere of subject-specific interest, for example a sewage works, farm or nature reserve, giving you the opportunity to see firsthand elements of the theme being taught. These visits may be led by an industrial host, an expert in the area or a member of academic staff. You may also be given a short task, or group work, to complete. If information on this is given out in advance, familiarise yourself with the material before the fieldtrip. Tips on how to complete the task may also be given during the introductory talk. With limited time available it is important that you hit the ground running.

General tips:

- Make sure you do all the preparatory reading indicated by the course leader; this will ensure maximum benefit is gained from the short time in the field.

- The fieldtrip may be your only opportunity to gain knowledge from an outside expert – keep up with him or her so you can hear what he or she is saying. Ask questions/clarify things you are unsure of.

- Make notes as you go along, or consider using a portable recording device and write up your notes soon after you return from the field.

- Make a note of experts' contact details or websites/references they recommend.

- While it's only a short trip, it's still fieldwork; make sure you take all necessary kit and precautions as indicated earlier.

- Make brief reflective notes after the visit. Did you find it interesting? What questions did the trip raise in your mind? Would you like to investigate this topic further?

4.10.2 Ongoing field investigations

Sometimes field investigations can continue over several days, weeks or even months. If your protocol requires repeated measurements to be taken over a period of time you will first need to consider whether this is realistic and sustainable. The protocol will need to account for natural variation that occurs over time, allowing you to consistently collect standardised data (especially if different individuals are involved). For example, variables such as time of day, weather conditions, point within a plant's life cycle may affect the data being collected. These variables must therefore be recorded at the same time you collect your data if they have not already been accounted for within your experimental design.

An extended sampling period will involve additional planning, commitment of time and further travel costs, and inevitably there is an increased risk that problems may arise. Therefore, as always, plan carefully and try to anticipate potential problems. Don't forget to pre-plan your statistical analysis to avoid any temporal pseudoreplication (Heffner et al., 1996).

General tips:

- If measuring the same point, plant etc. you need to mark it so you will easily find it again – what looks like an obvious marker in February may be hidden in summer when vegetation has grown.

- If you are working in a public area, keep markers discrete so inquisitive members of the public don't move them or damage the field area.

- Consider whether you should process samples as they are collected or whether you should store the samples and process them all together.

- Try to avoid working alone.

4.10.3   Residential trips – home and abroad

You should have a reasonable amount of time to plan for these; so use this time wisely. Residential trips are likely to provide an exciting part of your studies and may involve travelling to regions or countries you have not visited before. It is sensible to research where you are going and read the materials provided by tutors. Make sure you are clear about travel arrangements, the activities to be undertaken, preparatory work required, the assessment for the trip etc.

General tips:

- Be prepared – find out what to expect.

- Plan in advance: passports, visas, vaccinations, equipment sent ahead.

- Pack appropriate and spare clothes.

- Make sure you take adequate money.

- If you have special requirements, are overly nervous or have specific concerns about the trip, speak with your tutor or the trip convenor well in advance of departure date.

- Don't get too tired (partying!).

4.11   Overcoming the problems that WILL occur

When learning theory, in a lecture or tutorial, or practical techniques in a prescribed laboratory session, it is unlikely things will deviate drastically from what has been planned; if they do equipment is likely to be available and people will be on hand to help. In the field, however, it is likely you will have to solve 'real' problems, in a time-limited environment. These may be minor: a puncture on the minibus, cuts and bruises, ripping your wellies etc. and all of these can be easily solved if you are properly prepared. More serious problems might involve: dropping a vital piece of kit into deep water; the species under study going through its life cycle earlier/later in the year than usual; or disturbing a hornets' nest.

Other people can often cause difficulties, so make sure you know who you are travelling with, that everyone is well briefed and that you have exchanged contact phone numbers in case there is a change of plan. On arrival in the field it may become apparent that the careful plans you have made are completely unworkable; e.g. the species you were hoping to study no longer exists at that site or the field site is no longer accessible. In such circumstances it is important to reassess your goals. Is there another, similar species or nearby field site you could study? However, it is also important to be sensible and to realise when it is better to abandon the fieldwork and reschedule. If you do change your research plans you will need to contact other group members/ your tutor to discuss the change. Remember that good communication will ensure that the collective efforts are well coordinated.

Successfully overcoming unexpected problems can give you a huge confidence boost. If, for example, you work out a way of completing an investigation without a piece of equipment that has been broken, it will provide evidence of your problem-solving skills. These can then be highlighted on an application form for a job, or at an interview, and information of this nature will help you stand out from other candidates. In addition the experience will tell you a lot about yourself and how you cope in times of difficulties. This, in itself, can be one of the most important learning outcomes from conducting fieldwork. Chapters 1 and 2 of this book provide more information about problem solving.

4.11.1   Case study

You get to the field and the soil moisture probe you were planning to use is broken. What can you do? Consider the following:

- Is knowledge of soil moisture content critical to the research design or just interesting to know? If you don't know soil moisture content, can you reliably test your hypothesis or must you know soil moisture content to reach conclusions?

- Could you go back and get another probe and return the same day? Is there time to do this? Is there another probe you could use?

- Could you postpone the work and return another day? A possibility if there isn't enough time to go back the same day, or another moisture probe isn't available. However, does the planned research depend on the work being done on a particular date?

- Could you complete the rest of the fieldwork and return at a later date to measure soil moisture content? Does the soil moisture content need to be assessed at the same time as the other measurements, or do you just need to compare moisture content at different sites generally?

- Can you work out soil moisture content in another way? For example, take a soil sample, weigh the soil then dry it at $700\,°C$ for 72 hours, re-weigh the soil and calculate moisture content; do you have collection bags, somewhere to weigh and dry; do you have time for this alternative approach?

Can you think of an alternative solution to the problem?

4.11.2   Case study

Consider the following:

You get to a river and discover the water flow is very low, making it difficult to undertake the kick sampling[*] (Figure 4.4) planned. Try to think of a number of potential solutions to this problem. You should also refer to Chapter 1 for more ideas about solving problems creatively.

See Section 4.17 at the end of the chapter for a list of potential solutions.

---

[*] 'Kick sampling' is a standard method for streams and rivers that involves a three-minute 'kick' of the river bed upstream of a net, held on the stream/river bed, which catches invertebrates carried into it by the current.

Figure 4.4   Student group wading upstream to reach kick-sampling site

4.12   Feedback and assessment (see also Chapter 10)

As fieldwork is different from many of the learning activities you will undertake at university, tutors may use a range of interesting mechanisms to assess your knowledge and skills in the field. However, your tutor should notify you well in advance of how and when your field sessions will be assessed. If you are unclear, ask them.

You may get a great deal of informal feedback on your work during the field session. For instance your lecturer may move from group to group suggesting how a technique could be improved or may informally test you on your plant identification skills. Some tutors will carry out formal assessment in the field: for example by asking individuals to identify 10 plant species within a quadrat or by asking you to explain and demonstrate the use of a piece of equipment.

During residential trips, the evenings may be used for the preparation of materials that will be assessed. For example, you may be asked to prepare a short presentation, collate a photographic diary of the day's activities, write a short article on your findings for a magazine or to demonstrate the techniques you have learnt to others who were engaged in a different activity. Some of these assessments may 'count' (summative assessment, see page 195); others may be undertaken just to enhance learning and development of skills (formative assessment), but it is usually expected, and indeed part of the ethos of fieldtrips, that students should participate in all learning and assessment activities whether they count or not! Assessment will usually also take place after the field activity. This could include, for example, assessment of field reports, a field handbook or a research paper you have compiled based on your own work. It is essential that you are clear about the marking criteria for these items of work and know the deadline for submission. The fact that you may be expected to submit these assessments some time after a field session emphasises the need for clear protocols, and clear and thorough recording of information in a field notebook etc., so that you have easy access to all of the material you will need to complete the assignment.

4.12.1   Tutor notes

If the assessment methods you will be using are new to the students, take some time to explain the procedures so they know what is expected of them. They may be apprehensive about being assessed in an unfamiliar way, so you could use a similar formative assessment procedure prior to the main summative assessment. Consider the accessibility of the method you are using.

Fieldwork studies should provide good opportunities for feedback as students carry out their work. However, students don't always recognise they are receiving feedback when you chat to them in the field. Remember to highlight that what you are doing is providing feedback!

4.13   Concluding comments

Fieldwork is an important and rewarding part of many bioscience degree courses. Though potentially hard work, fieldwork provides opportunities to gain friendships, a greater understanding of your subject, key research skills and many transferable skills including: communication; problem solving; project management; time management; team work; leadership; organisation; safety awareness; and independent thinking. This chapter will hopefully help you make the most of your fieldwork experiences.

4.14   How you can achieve your potential during fieldwork

- Carry out a thorough risk assessment and work safely at all times.

- Understand the importance of planning and time management before, during and after fieldwork.

- Ensure the Scientific Method is applied during fieldwork.

- Be prepared to interact with others and work as part of a team to solve problems.

- Understand how technology can enhance learning at all stages during fieldwork.

- Understand the methods used by your university to assess fieldwork.

- Be aware of the ethical implications and financial costs of conducting fieldwork.

4.15   References

Andrews, J., Kneale, P., Sougnez, W., Stewart, M. and Stott, T. (2003) Carrying out pedagogic research into the constructive alignment of fieldwork. *Planet* **11**, 51. www.gees.ac.uk/pubs/planet/pse5back.pdf.

Baggott, G.K. and Rayne, R.C. (2007) The use of computer-based assessments in a field biology module. *Bioscience Education* **9**, 5. www.bioscience.heacademy.ac.uk/journal/vol9/beej-9-5.aspx.

Baggott, G.K. (2009) Fieldwork: e-learning benefits the part-time student. *UK Centre for Bioscience Bulletin* **26**, 3. Available at: www.bioscience.heacademy.ac.uk/resources/themes/fieldwork.aspx. Accessed June 2010.

Birnie, J. and Grant, A. (2001) Providing Learning Support for Students with Mental Health Difficulties Undertaking Fieldwork and Related Activities. Geography Discipline Network, University of Gloucestershire (http://www2.glos.ac.uk/gdn/disabil/mental/index.htm). Accessed April 2011.

Boyle, A., Maguire, S., Martin, A. *et al.* (2007) Fieldwork is good: the student perception and the affective domain, *Journal of Geography in Higher Education* **31**, 299–317.

Chalkley, B. and Waterfield, J. (2001) Providing Learning Support for Students with Hidden Disabilities and Dyslexia Undertaking Fieldwork and Related Activities. Geography Discipline Network, University of Gloucestershire (http://www2.glos.ac.uk/gdn/disabil/hidden/index.htm). Accessed April 2011.

Cotton, D.R.E. and Cotton, P.A. (2009) Field biology experiences of undergraduate students: the impact of novelty space. *Journal of Biological Education* **43**, 169–174.

Dytham, C. (2003) *Choosing and Using Statistics: A Biologist's Guide* (2nd Edition). Oxford: Blackwell Science.

Gold, J.R., Jenkins, A., Lee, R. *et al.* (1991) Fieldwork. In *Teaching Geography in Higher Education – a Manual of Good Practice*, Ed. J.R. Gold, A. Jenkins, R. Lee *et al.* Oxford: Blackwell Science, pp. 21–35.

Healey, M., Jenkins, A., Leach, J. and Roberts, C. (2001) Issues in Providing Learning Support for Disabled Students Undertaking Fieldwork and Related Activities. Geography Discipline Network, University of Gloucestershire (http://www2.glos.ac.uk/gdn/disabil/overview/index.htm). Accessed April 2011.

Heffner, R.A., Butler, M.J. and Keelan-Reilly, C. (1996) Pseudoreplication revisited. *Ecology* **77**, 2558–2562.

Kent, M., Gilbertson, D.D. and Hunt, C.O. (1997) Fieldwork in geography teaching: a critical review of the literature and approaches. *Journal of Geography in Higher Education* **21**, 313–332.

Nairn, K., Higgett, D. and Vannests, D. (2000) International perspectives on field courses. *Journal of Geography in Higher Education* **24**, 139–149.

Maskall, J., Stokes, A. Truscott, J.B. *et al.* (2007) Supporting fieldwork using information technology. *Planet* **18**, 18–21. www.gees.ac.uk/planet/p18/jm.pdf.

Maskall, J. and Stokes, A. (2008) *GEES Subject Centre Teaching and Learning Guide: Designing Effective Fieldwork for the Environmental and Natural Sciences.* Plymouth: The Higher Education Academy Subject Centre for Geography, Earth and Environmental Sciences. Available at: www.gees.ac.uk/pubs/guides/fw2/GEESfw-Guide.pdf. Accessed May 2010.

Maw, S.J., Mauchline, A.L. and Park, J.R. (2011) Biological fieldwork provision in Higher Education. *Bioscience Education* **17**, 1. www.bioscience.heacademy.ac.uk/journal/vol17/beej-17-1.aspx.

4.16 Additional resources

4.16.1 *General*

Jenkins, A. (1994) Thirteen ways of doing fieldwork with large classes/more students. *Journal of Geography in Higher Education* **18**, 143–154.

Jones, A., Reed, R. and Weyers, J. (2007) *Practical Skills in Biology.* Harlow: Pearson Education.

Scott, G. and Goulder, R. (2009) Making the most of fieldwork: doing less to achieve more! *UK Centre for Bioscience Bulletin* **27**, 10. www.bioscience.heacademy.ac.uk/events/themes/fieldwork.aspx.

Southwood, T.R. and Henderson, P.A. (2000) *Ecological Methods* (3rd Edition). Oxford: Blackwell Scientific. There is also a website that provides additional material to support the book: www.blackwellpublishing.com/southwood. Accessed April 2011.

Wilmot, A. How to … Make Your *Field Trips* More Sustainable (www.bioscience.heacademy.ac.uk/ftp/esd/howtofieldtrips.pdf). Accessed April 2011.

The Field Studies Council is an environmental education charity committed to helping people understand and be inspired by the natural world. Their field study guides are available from the publications section of their website, www.field-studies-council.org. Accessed April 2011.

The Young Explorers Trust and the British Ecological Society have published a guide to fieldwork for youth expeditions. Although this guide isn't written for undergraduates, the *Techniques and Tools* section may be useful. www.britishecologicalsociety.org/educational/fieldwork/yet_exploring_a_changing_world.php. Accessed April 2011.

The Engage in Research website has been designed to help biology undergraduates with their research projects. Advice covers all aspects of completing a project from planning to statistics and report writing. www.engageinresearch.ac.uk. Accessed April 2011.

Learn higher: Group Work Resource Home. This site is designed to help both students and tutors understand and overcome the challenges of group work. www.learnhighergroupwork.com. Accessed April 2011.

Free Fieldwork – What's the Story? Leicester University have compiled a list of questions to help you work out the full cost of fieldwork as offered by different Universities. http://www2.le.ac.uk/departments/geology/undergraduate-courses/free-fieldwork-whats-the-story. Accessed April 2011.

The Countryside and Rights of Way Act (2000): www.opsi.gov.uk/acts/acts2000/ukpga_20000037_en_1. Accessed April 2011.
The Scottish Outdoor Access Code: www.outdooraccess-scotland.com. Accessed April 2011.

4.16.2   Safety (accessed April 2011)

Winsor, S. And France, D. (2008) Fieldwork Safety: A Resource Guide Briefing on BS8848. Plymouth: The Higher Education Academy Subject Centre for Geography, Earth and Environmental Sciences (www.gees.ac.uk/pubs/ other/BS8848BriefFeb2608.doc).
Aldiss, D.T. (2007) NERC Guidance Note: A Safe System of Fieldwork. Swindon: Natural Environment Research Council. Provides more detailed safety information. www.nerc.ac.uk/about/work/policy/safety/documents/ guidance_fieldwork.pdf.

4.16.3   Data handling

Dytham, C. (2003) Choosing and Using Statistics: A Biologist's Guide (2nd Edition). Oxford: Blackwell Science.
Fowler, J., Cohen, L. and Jarvis, P. (1998) Practical Statistics for Field Biology (2nd Edition). Chichester: John Wiley & Sons, Ltd.
Hurlbert, S.H. (1984) Pseudoreplication and the design of ecological field experiments. Ecological Monographs (Ecological Society of America) 54, 187–211.
Kokko, H. (2007) Modelling for Field Biologists and Other Interested People. Cambridge: Cambridge University Press.
Sutherland, W.J. (2006) Ecological Census Techniques: A Handbook (2nd Edition). Cambridge: Cambridge University Press.

4.16.4   Accessibility (accessed April 2011)

Six web-based guides in the series Learning Support for Disabled Students Undertaking Fieldwork and Related Activities. Geography Discipline Network, University of Gloucestershire. Available at: http://www2.glos.ac. uk/gdn/disabil/.
Equality Act 2010: Implications for Higher Education Institutions: www.ecu.ac.uk/publications/equality-act-2010.

4.16.5   Resources for tutors (accessed April 2011)

The Fieldwork pages of the UK Centre for Bioscience website contain links to articles and reports on teaching fieldwork in Higher Education. www.bioscience.heacademy.ac.uk/resources/themes/fieldwork.aspx.
The British Ecological Society website provides university teaching resources for fieldwork. www.britisheco-logicalsociety.org/educational/teaching_resources/university_resources.php.
Supporting Fieldwork Using Information Technology conference abstracts: www.gees.ac.uk/projtheme/cetls/el/ fwitconf06.htm.

4.17   Potential solutions for kick-sampling case study

- Undertake kick sample anyway, but be aware of how the low water flow could affect your results.
- Go to a different site where the water flow is greater? Do you know of a suitable alternative site? Could you get there in time? Do you have permission to work there? What was the importance of your chosen site: will it affect your research to move sites?
- Postpone the work and return at a later date?
- Use a different sampling technique?

5 *In vivo* work

David I. Lewis

5.1 Introduction

In vivo studies are studies of biological processes or experiments occurring, or carried out, in the living organism; *in vivo* being derived from the Latin *vivus* meaning '*(with)in the living*'. In contrast, *in vitro* studies or experiments are those undertaken under artificial conditions outside of a living body, e.g. in a test tube, culture dish or isolated tissue bath.

In vivo studies are not a recent development in the process of scientific discovery; among the earliest records of studies on living animals are the experiments of the Greek philosopher, Aristotle, in the fourth century BC. However, the title of the 'Father of Vivisection' goes to the Roman physician Galen (second century AD) for his studies in live pigs. *In vivo* studies have remained at the forefront of advances in medicine ever since, the number and complexity of such studies mushrooming with the rapid developments in scientific discovery over the last 150 years.

Following the sequencing of the human genome, with the potential this may provide for the development of new medicines, and the parallel development of powerful new molecular biological technologies and techniques, there has been a marked reduction in the use of intact animals or isolated animal tissues in scientific and medical research. However, over the last few years, it has become apparent that research using *in vivo* animal models, or substantially intact organ systems, is essential for scientific progress. Organ systems and organisms are much more than the sum of their individual parts; investigating mechanisms at the cellular and sub-cellular level will not allow us to fully understand how individual organs or indeed whole bodies function.

As a consequence of this reductionist approach (investigating physiological function at a cellular or sub-cellular level rather than in an intact organ or animal) to scientific research adopted over the last 20 years, there is now a global shortage of individuals with the necessary expertise and skills required to undertake whole animal or *in vivo* studies. There is also a shortage of individuals who are capable of training the next generation of *in vivo* scientists. In 2004, a survey conducted jointly by the British Pharmacological Society and the Physiological Society estimated that 25% of university lecturers who teach integrative physiological and pharmacological techniques to students were due to retire within five years (British Pharmacological Society and The Physiological Society, 2006). As a consequence, significant initiatives have been put in place to address this skills shortage. In the UK, major pharmaceutical companies have established the 'Integrative Pharmacology' Fund to promote research and teaching in *in vivo* studies; in collaboration with government and other

Effective Learning in the Life Sciences: How Students Can Achieve Their Full Potential, First Edition.
Edited by David J. Adams.
© 2011 John Wiley & Sons, Ltd. Published 2011 by John Wiley & Sons, Ltd.

partners, they have committed £12M over five years to establish four Centres of Excellence in Integrative Mammalian Biology. In the USA, the National Institute of General Medical Sciences has begun funding short courses in Integrative and Organ System Pharmacology, and the American Society for Pharmacology and Experimental Therapeutics has established the ASPET Integrative Organ System Science Fund to provide enhanced training in integrative and organ system science. The International Union of Pharmacologists (IUPHAR) has a similar initiative to provide training in countries where it is not currently available.

There are other key skills in addition to technical expertise that are required by *in vivo* scientists. An Association of the British Pharmaceutical Industry and UK Biosciences Federation (2007) report, *In Vivo Sciences in the UK: Sustaining the Supply of Skills in the 21st Century*, highlighted these key skills as:

- Knowledge of the relevant legislation, the regulatory framework and the responsibilities of individuals.

- Knowledge and understanding of ethical review processes.

- Knowledge and understanding of the principles of the 3Rs.

- Knowledge and understanding of best practice in animal welfare and husbandry.

- Knowledge of animal models of disease.

- The ability to recognise both normal animal behaviours and the signs of ill health, pain or distress.

- Training in experimental and statistical design with respect to animal studies.

- The ability to conduct harm/benefit analyses.

- Knowledge and understanding of the alternatives to the use of animals or animal tissues in scientific studies.

- Knowledge and understanding of the key arguments for and against the use of animals in scientific research.

Furthermore, it is essential that training in these key skills is provided early in an individual's career. This chapter will address and provide examples of training in these important areas.

5.2   Animal welfare legislation

The stringency of animal welfare legislation varies considerably across the world, with substantial differences in the protection afforded to animals used for research in different countries.

5.2.1   *Case study*

You are a member of the legislature in your own country, tasked with drawing up a new animal welfare law designed to protect animals used for research. What would you include in this legislation? In designing your own legislation, you need to balance the scientific requirements against the welfare of any animal being used for this research. Things that you might like to consider are:

- Which animals should be protected? Would this protection extend to invertebrates?

- The severity of permitted procedures. How would you ensure that the pain, suffering or distress experienced by the animals used is minimised?

- Animal husbandry. How would you house and care for these animals?

- How would you incorporate the ethical review of any studies? For example, do the benefits to science or medicine outweigh the harm to experimental animals? Have the principles of the 3Rs (see below) been applied? Is best practice in animal welfare being promoted? Who will undertake the review process: an institution (e.g. university) or a government agency? (For more detailed consideration of ethical issues in the biosciences, see Chapter 9.)

- How would you monitor work with animals and ensure compliance with the legislation?

5.2.2 Tutor notes

Following completion of the case study by students, you could discuss their suggestions for items to be included in the law with respect to current animal welfare legislation, both within your country and with best practice elsewhere in the world. This discussion could include: should all species (including invertebrates) be protected? Should higher-order species (e.g. primates) be afforded greater protection? How would you promote the alternatives to the use of animals? Should there be restrictions on the severity of individual procedures and how would you minimise pain, suffering or distress? How would you ensure best practice in animal husbandry, balancing the scientific requirements against meeting the physiological and psychological needs of the animals (cage sizes, environmental enrichment, group housing etc.)? Would you permit animals to be used multiple times for the same procedure (re-use) or be used in more than one procedure (continued use)? What ethical review procedures would you incorporate into the legislation? What training and licences would you require for research scientists? Who would monitor compliance and what penalties should there be for breaking the law? Should animals culled in order to provide animal tissues be similarly protected and how would you ensure that all animals are humanely culled?

5.2.3 Current legislation

In the United States, scientific studies using animals are regulated by the Animal Welfare Act 1966 (and subsequent amendments) (US Department of Agriculture, 1966; see Appendix). The Act, however, only protects certain species, e.g. non-human primates, dogs, guinea pigs and cats; it specifically excludes birds, fish, invertebrates, rats and mice, the last two accounting for over 85% of animals used for research. In addition to Federal regulations, institutions using animals in research, teaching or testing may be subject to additional State and local laws which provide additional protection.

Within the European Union, scientific and medical research involving the use of animals is governed by the European Union Directive 86/609/EEC (European Union, 1986). The aims of this directive are to harmonise, across the European Union, laws covering the welfare of animals used in scientific research, and to set minimum standards of animal care and welfare. Unlike the US Animal Welfare Act, Directive 86/609/EEC covers all species of non-human vertebrates in any research studies that may cause the animals pain, suffering, distress or lasting harm. Following publication of the directive, each European country then enacted laws to incorporate it into their national legislation. In the United Kingdom the relevant legislation was the Animal (Scientific Procedures) Act, 1986 (HM Government, 1986). A(SP)A is widely regarded as the strictest piece of animal welfare legislation in the world, far exceeding the minimum requirements of the directive. Its stringency is the result of the requirement to apply the principles of the 3Rs (replacement, refinement and reduction of the use of animals in research – see below), not only in the licence application but also on a day-to-day basis throughout the lifetime of the project.

In September 2010, the European Union adopted Directive 2010/63/EU as a replacement for the existing Directive 86/609/EEC. This new directive is considerably more detailed and more prescriptive: it specifies much more explicitly what is and is not permitted, for example, there are now defined permissible purposes; it imposes greater restrictions on which species can be utilised (including fetal and larval forms), requires the principles of the 3Rs to be fully implemented, and requires each facility or institution which breeds, holds or uses animals to establish an 'Animal Welfare Body'.

5.3   The principles of the 3Rs

The principles of the 3Rs were first introduced by Russell and Birch (Russell and Birch, 1959); their application provides an ethical framework by which scientific research involving animals can be undertaken humanely. The principles are:

- **Replacement:** of 'protected' animals (all living vertebrates except humans) used in scientific research with alternative techniques.

- **Refinement:** of scientific procedures to minimise animal pain, suffering or distress.

- **Reduction:** of the number of animals used by obtaining more scientific data from the same number of animals or the same amount of data from fewer animals.

5.3.1   Replacement

Elimination of the need to use animals in scientific research is the ultimate goal of all individuals whose research involves their use. Replacement can be absolute: to replace the use of animals with alternative techniques or materials that don't involve animals or animal tissues at any point, for example computer modelling of body systems, *in silico* methods or the use of human material (e.g. obtained during surgical procedures). Alternatively, replacement can be relative, using techniques or systems that do not involve the use of live animals but use animal tissues at some point, e.g. established cell cultures of animal tissues, animal tissues obtained from abattoirs, animal tissues obtained from animals that have been culled humanely solely to provide this tissue, or the use of invertebrates, e.g. zebra fish, flies or the nematode, *Caenorhabditis elegans*.

5.3.2   Refinement

Refinement can be brought about by changing procedures or techniques so that the pain, suffering or distress experienced by an animal is minimised, not just for the duration of the experimental procedure but over its entire life. Refinement could include the administration of analgesics or anaesthetics (where appropriate) to alleviate pain; training animals with food rewards to cooperate with experimental procedures, e.g. training a monkey to put its arm out so a blood sample can be taken; or the use of non-invasive techniques such as imaging or telemetry (the remote radio monitoring of physiological parameters, including blood pressure, heart rate and nerve activity, using implants, thereby removing the need to capture or restrain animals in order for these measurements to be taken).

Substantial improvements in animal welfare can be achieved by good animal husbandry, housing animals used for research in as near natural conditions as possible. Housing them in such conditions enables them to express natural behaviours, improving the quality of their lives whilst minimising physiological and mental stress. Not only does this benefit the animals, it also improves the quality of the research data obtained. Consider an animal in a zoo that spends all its time pacing up and down along the front of its enclosure. Such an animal would be psychologically stressed and, as a result, would have an altered physiology and biochemistry. If it were an animal which was going to be used for research, any data obtained from it would be of limited value and may vary considerably from data obtained from non-stressed animals.

5.3.2.1 Case study

Choose an animal that is used in scientific research (e.g. rat, dog, monkey or rabbit). Consider how it lives in the wild. Now design an enclosure or cage to keep this animal in an animal facility or laboratory. You need to balance the requirements of the science or scientist against good animal husbandry and meeting the physiological and psychological needs of the animal. Things you may want to consider are:

- Should the animal be housed individually or as part of a group? If the latter, the size of the group.

- The size and design of the enclosure.

- Access to food and water.

- Environmental enrichment: items that provide physiological and psychological stimulation for the animal.

5.3.2.2 Tutor notes

Allocate students to groups and ask each group to design an enclosure for their chosen animal; they should take into account how the animal lives in the wild. On completion, get each group to present their design to the class, explaining why they have included each element. Discuss what they have included compared to what occurs in practice. Images and videos of enclosures/cages for different animals used in research can be downloaded from the 'Understanding Animal Research' website (www.understandinganimalresearch.org.uk/homepage).

5.3.3 Reduction

If the use of animals is unavoidable, then measures should be taken to reduce the number of animals required to a minimum. This can be achieved in a number of ways, including good experimental and statistical design, so that the optimal number of animals is used, as the use of too few animals (which may lead to results that are statistically insignificant) is just as bad as using too many animals. Power calculations (statistical methods for determining whether the optimal number of animals is being used; http://invivostat.co.uk/) are one way of estimating the appropriate number of animals. Alternatively, the numbers of animals required could be reduced by taking repeated measurements from the same animals (e.g. using modern technologies such as imaging or telemetry), by measuring more variables in fewer animals or by using genetically similar (i.e. in-bred) animals. With genetically similar animals, there is less variability in any data and therefore fewer animals are required. Animal numbers can also be reduced by the sharing of animals, particularly when culling animals to provide tissue for experimental studies.

5.4 Alternatives to the use of animals in the development of new medicines

In the 'replacement' section above, I discussed potential alternatives to the use of animals in scientific and medical research. The question then arises as to whether we need to use animals at all in research that leads to the development of new medicines. Can we replace them with alternatives such as computer modelling or non-animal, cell-culture-based assays? Unfortunately, despite the hundreds of millions of pounds spent each year by industry and government in developing these alternative methods or techniques, at present they cannot fully replace animals. However they can significantly reduce the number of animals required as they can be used instead of animals as high-throughput screens in the early stages of the drug development process. This can also lead to significant savings in costs and time.

We currently have computer models of individual organs or tissues, e.g. heart, uterus or parts of the brain, but not whole bodies. Computer models are only as good as our current knowledge of these individual organs, which in many cases is extremely limited. They also require significant computing power. For example, scientists at the IBM Almaden Research Laboratories and the University of Nevada ran a simulation of half a mouse brain on the BlueGene L supercomputer (Frye *et. al.*, 2007). In spite of this machine having 4 096 processors, each one of which used 256 MB of memory, the vast complexity of the simulation meant that it only ran for 10 seconds at a speed 10 times slower than real life, equivalent to 1 second in the life of a mouse brain. This mouse simulation involved 8 million neurons each with up to 6 300 synapses; only a fraction of what would be required if we were to attempt to emulate the whole human brain, which is estimated to contain 100 billion neurons, each with between 7 000 and 10 000 synapses.

Whilst we may not be able to fully model individual organs or tissues, let alone whole bodies, computer modelling plays a vital role in the initial stages of the drug discovery process. Typically, when developing a new medicine, a pharmaceutical company will start off with 10 000 very similar compounds, each potentially the next cure for the particular disease under investigation. Most medicines act by binding to receptors located on the surfaces of cells or by modulating the activity of intracellular enzymes. We have the capability to model these individual target sites and therefore can use computer modelling as a high-throughput screen to select the drug candidates which provide a 'best fit' with the active sites of these receptors or enzymes.

In the next stage of the drug development process, cell-based assays can be utilised to look at drug modulation of intracellular processes or as a replacement for animals in investigations into whether a new medicine is mutagenic or otherwise harmful to cells or tissues. The 'Draize test' was a commonly utilised safety test to determine whether a compound caused skin or eye irritation. In this test, the compound was topically applied either to the eyes of rabbits or to an area of shaved skin on their backs to determine whether it caused eye or skin irritation respectively. By their very nature, these tests caused significant pain and distress to the rabbits and there was therefore a significant impetus to develop non-animal-based alternatives. This led to the development of Episkin (www. skinethic.com/EPISKIN.asp), a human keratinocyte cell culture assay which can determine whether test agents cause tissue damage, the production of inflammatory mediators or cell death. Episkin has now been accepted by the European Centre for the Validation of Alternative Methods Scientific Advisory Committee (ECVAM, 2007) as an approved and valid alternative to the Draize skin irritation test. Similarly, the bacteria-based 'Ames test' can now be used instead of rats or mice to determine whether a chemical is likely to damage DNA or has the potential to cause cancer. Isolated organ or tissue preparations can then be utilised to investigate the effects of putative new medicines on individual organs or tissues, e.g. the Langendorff isolated perfused heart preparation, which enables the determination of drug effects on cardiac contractile strength (inotropic effects), heart rate (chronotropic effects) and vascular effects without the neuronal and hormonal complications of an intact animal model. Brain slice preparations can be used to investigate effects of experimental compounds on neuronal firing rate, receptors or ion channels.

However, we take most of our medicines orally. The medicine is then absorbed from the stomach and passes via the hepatic portal vein to the liver where most ingested compounds are metabolised to active agents. Some of this active medicine will be lost into the bile; the remainder will pass into the blood. Will it be of a therapeutic concentration? Will it be able to get to the target organ(s) or tissues? Will it have adverse effects on other organs or tissues? How quickly will it be lost from the blood? All of these questions can only be addressed in an intact system. Whilst 'alternative' techniques or technologies cannot provide the answer to these questions, their use as high-throughput screens can, by filtering out unsuitable test compounds, reduce the number of compounds that need to be tested in animals from, for example, an original 10000 to approximately 10; their use therefore will substantially reduce the number of animals required.

If, ultimately, we have to test our putative new medicine in an intact system, why should we use animals? If the medicines are for human use, surely they should be tested on humans? But would you want to be the first person to use a new medicine that had not previously been tested in animals, not knowing what the side effects might be or, indeed, whether it would work at all? An alternative might be to test a medicine on people that some may consider to be of limited use to Society – individuals in a permanent vegetative state or prisoners on death row? There are plenty of precedents from history where we have trialled new medicines or investigated disease processes in individuals without their consent or knowledge. For example, the Nazi concentration camp studies, the Tuskegee syphilis studies on African-American sharecroppers (Centers for Disease Control and Prevention, 2011) or the studies undertaken by the US in Guatemala (US Department of Health and Human Services, 2011) in which prisoners and others incapable of giving fully informed consent were deliberately infected with sexually transmitted diseases. Fortunately, the European Union Convention on Human Rights and Biomedicine (Council of Europe, 2007) prohibits studies on individuals without their consent. It also prohibits studies on vulnerable individuals (e.g. children, those with mental incapacity, or prisoners) if equivalent studies can be carried out using individuals who have the capacity to give fully informed consent. The Office for Human Research Protections (www.hhs.gov/ohrp/) provides similar safeguards in the United States.

If medical tests involving individuals who have either not given their consent or are deemed vulnerable are wrong, what right do we have to test new medicines on animals who equally cannot give their consent? The answer lies with an individual's conscience. However, in support of the use of animals in research, we are not using them solely for our benefit. Nine out of ten medicines used to treat or prevent human disease are also used in veterinary medicine. Animals are also used to develop medicines used only to treat animal diseases, e.g. the development of a vaccine against feline leukaemia, a disease caused by a virus, which is often fatal in cats.

5.5   Animal models of disease

There are many examples of medicines having different, often toxic, effects in animals and no effect in humans or vice versa; e.g. paracetamol is fatal in cats because they lack the glucuronyl transferase enzymes required to safely break it down, whilst arsenic is toxic to humans but has no effect in sheep. Similarly, medicines that have been pronounced safe as the result of animal experiments e.g. TGN1412 (Attarwala, 2010) and thalidomide (Botting, 2002) cause severe (or fatal in the case of TGN1412) side effects in humans. Therefore, can animals really be used to safely test medicines for human use or are we only using animals because there is no viable alternative?

We share the same basic building blocks; a mouse has 98% genetic similarity to a human. We also have the same organs (e.g. heart, lungs, brain, kidneys) in the same places doing the same things. Animals also suffer from the same diseases as us: obesity, diabetes, cancer, heart and neurological diseases to name but a few. We can use animals with these diseases to increase our understanding of the disorders and develop new treatments for them.

The selective breeding of animals used for research, spontaneously occurring genetic mutations and the development of genetically modified animals have enabled us to develop animals that more closely model human diseases for use in research. If two individuals eat the same amount of food, they will increase their body weights by different amounts; the same applies to rats. Selective breeding of both high- and low-body-weight-gain animals has resulted in the generation of two colonies of animals: diet-induced obese (DIO) animals which will become obese in adulthood, and diet-resistant (DR) animals which exhibit minimal increases in body weight (Madsen *et al.*, 2010). Similar selective breeding has resulted in the generation of a population of rats which will develop spontaneous hypertension (SHR) in adulthood. Spontaneous mutations occurring in breeding colonies of animals used for research have resulted in other models of human diseases, e.g. the Ob/Ob mouse which lacks leptin, a key circulating adiposity signal, the consequence of which is that these animals become obese in adulthood.

Following the sequencing of both the mouse and human genome, we can now knock out individual genes to generate genetically modified animals which have diseases or disorders caused by defects in these particular genes. A good example is the cystic fibrosis mouse, which lacks the cystic fibrosis transmembrane conductance regulator (CFTR) gene and hence has abnormal functioning of the ion channels which transport chloride across lung epithelial membranes. Whilst this technology is suitable for creating animal models where the underlying defect is due to a single gene mutation, most of the diseases or disorders affecting humans are multifaceted, involving a multi-gene susceptibility and other factors (e.g. environment). In such cases, and especially where we have not identified all of the factors that contribute to the disease, we can only generate animal models which mimic elements of the disease pathology. An example is the PDAPP mouse, a transgenic model of Alzheimer's disease which exhibits some of the pathological hallmarks of the disease, including neuritic plaques, but not neurofibrillary tangles or clear signs of neurodegeneration.

Animal models of human disease are therefore not perfect and, as a consequence, some medicines that were deemed safe based on the results of animal experiments have subsequently caused severe or even fatal side effects in humans. This was, in part, due to the procedures in place at the time to evaluate the safety of medicines. In the 1950s and 1960s, thalidomide, an anti-emetic, was given to pregnant women to combat morning sickness (Botting, 2002). However its use resulted in children being born with limb deformities. The reason the teratogenic effects of thalidomide were not picked up by animal studies was because, at that time, the safety of new medicines was only tested in adult male animals. As a consequence, procedures have been changed and the mutagenic and teratogenic potential of a putative new medicine is now assessed both in pregnant female animals and on males prior to conception (the latter in case it has effects on spermatogenesis or sperm motility). TGN1412, developed to mitigate the effects of autoimmune and immunodeficiency disease (Attarwala, 2010), was deemed safe following animal studies including those undertaken in monkeys. When given to human volunteers, it evoked a massive immune response and immediate multi-organ failure. Unlike most of our medicines to date, which are small molecules, TGN1412 is a human monoclonal antibody, the first to be used as a potential medicine. Its side effect, that involved the development of a severe immune response, would not have been picked up in *in vitro* systems or in other species. To prevent a similar tragedy occurring in the future, new medicines with such novel mechanisms of action are tested by 'micro dosing' (starting off with a low dose that is virtually physiologically undetectable) in a single individual with the appropriate medical support facilities (e.g. intensive care bed) available should anything untoward occur.

As indicated earlier, there are known differences between humans and other species; for example, the presence or absence of an enzyme required to metabolise a particular compound. This knowledge is utilised in determining which species to use in the evaluation of the safety of new medicines. The testing process is aided by the fact that many of our new medicines are so-called *me-too* compounds; that is they are very similar to medicines already on the market; the same species of animal is therefore used to test these structurally related compounds.

5.6 Experimental design

Good experimental design is essential, not only to ensure that the aims and objectives of a particular series of experiments are met, but also, if there is no alternative to the use of animals, that the optimal numbers of these are utilised.

5.6.1 Case study

Tom, a teenager paralysed from the neck downwards as the result of a car accident, is totally dependent on his carers to cater for his everyday needs. You are a research scientist trying to develop a cure for the spinal cord injury that afflicts Tom and others like him. Your laboratory has developed a rat model of spinal cord injury in which you inject a viral neural toxin into the lower spinal cord. Following the injection, the animals lose control of their lower limbs but are otherwise healthy; they are able to drag themselves around their cage using their upper limbs. You now wish to use this experimental model to test the efficacy of a potential new treatment for spinal cord injury that involves the injection of embryonic stem cells into the spinal cord. Pilot studies have shown that, following injection of the stem cells into the animal's spinal cord, they partially recover use of their lower limbs enabling them to move freely around the cage.

Design a series of experimental procedures that, collectively, will validate this potential new treatment for spinal cord injury: include *in vitro*, *in vivo* and human studies as appropriate. For the *in vivo* studies, consider specifically the experimental design issues that such studies involve and how you will include the 3Rs in your approach.

5.6.2 Tutor notes

Allocate students to groups and give them sufficient time to design a series of experiments to validate this new treatment for spinal cord injury. On completion, get each group to present their proposed series of experiments to the whole class. Ensure that *in vitro*, *in vivo* and human studies are covered, comprising both anatomical and functional aspects. With respect to *in vivo* studies, discussions could focus on: how you would assess return of function; electrophysiological and morphological studies; the implementation of the 3Rs, specifically minimising pain and suffering; animal husbandry; and power calculations to ensure the appropriate number of animals have been utilised. For the *in vitro* studies, you could refer students to databases of alternative techniques and technologies e.g. the ALTWEB (http://altweb.jhsph.edu/) and ALTBIB (http://toxnet.nlm.nih.gov/altbib.html) databases (see Section 5.13, Additional resources).

5.7 Recognition of pain, suffering or ill health in animals used for research

It is both the legal and also the moral responsibility of anybody who uses animals in research to minimise any suffering experienced by these animals. When an animal is in pain or distress, it will alter its natural behaviours: appearance, how it moves around etc. In order to recognise abnormal behaviour, you need to be fully cognisant of that particular species' pattern of normal or natural behaviours; these vary greatly between species. As an example, the normal behaviours of a rat and those exhibited when it is in pain or distress are compared in Table 5.1.

Table 5.1 Normal behaviour of a rat compared with the behaviour of a rat in pain or distress

Normal behaviour	Behaviour whilst in pain or distress
Gregarious	Isolated
	Crouched with head in abdomen
Active, alert, inquisitive	Reduced movement
	May stagger, circle or be ataxic
Avoids capture or handling	Excessive vocalisation when restrained
Warm, good muscle tone and well groomed	Poor coat condition and weight loss
	Piloerection of fur
	Reduced grooming
	Excessive scratching, biting or licking of affected area
Regular abdominal respiration	Irregular respiration
Normal food and water intake	Reduced food and water intake
	May be dehydrated (skin remains retracted when pinched)
Regular voiding of waste products	Reduced or no voiding of waste products

5.8 Ethical review of *in vivo* studies

In the United States, the Animal Welfare Act (1966) requires organisations that use animals for research to establish Institutional Animal Care and Use Committees (IACUCs) to provide ethical review of animal care and use within their institutions (US Department of Agriculture, 1966, 2009). In the United Kingdom, from 1997, institutions using, holding or breeding protected animals are required to set up an ethical review process (ERP) to provide local ethical review within the establishment to supplement central ethical review provided by the Government Animals in Scientific Procedures Inspectorate (Home Office, 1998). The remit of an ERP in the UK is to provide institutions with advice and support to facilitate high standards of animal welfare and the best science. It is called a process rather than a committee because it is involved before, during and after a programme of work; it provides input at the planning stage which continues once the work is in progress, and reflects on the lessons learned once the work has been completed. Membership of the ERP is flexible but has to include the Named Veterinary Surgeon at the establishment, animal care and welfare officers, scientists and lay members who have no responsibilities under the Animal (Scientific Procedures) Act or links to the institution.

The responsibilities of the ERP are:

- Promoting the development and uptake of the 3Rs.

- Examining applications for new and amended project licences, considering the balance of the likely costs to animals against the expected benefits of the work (harm/benefit analysis).

- Providing a forum for the discussion of the issues relating to the use of animals (updating staff on ethical advice, best practice and the law).

- Undertaking retrospective reviews of projects and continuing to apply the 3Rs to all projects throughout their lifespan.

- Considering the care and husbandry of all animals including breeding stock and the humane killing of animals.

- Regular reviews of the managerial systems and procedures where these impact on animal welfare.

- Advising on staff training and ensuring competence.

5.8.1 Case study

The aims of this case study are to provide you with a greater understanding of the Ethical Review Process (ERP) and to encourage you to consider your own views on what is or is not morally acceptable in scientific research involving the use of animals. Read the case study then consider, with some of your fellow students, the questions that follow; these are questions that an ERP may have to consider.

A study of the predator–prey response

This study will show whether the response of prey animals to a predator is a learnt behaviour from parents, an innate inherited response, or a response that only develops on exposure of the individual to a predator.

It is proposed to use three groups of mice with different origins; namely (1) wild caught mice; (2) the first generation of mice bred from captive wild mice; and (3) laboratory-bred mice obtained from suppliers licensed under the 1986 Animal (Scientific Procedures) Act. These three categories will contain mice with different experience of predators. Group (3) mice are very inbred animals which for many generations have had no exposure to predators. They are unlikely to respond in predator–prey interactions in the same manner as mice in the wild. They will serve as controls for the wild-caught mice. The first generation offspring of the wild-caught mice will themselves have had no exposure to predators but may have inherited learned avoidance responses from their parents, whereas the wild-caught mice will have developed, by experience, avoidance responses to predators.

Mice are the animal of choice because they are cheap and a larger number can be used compared to larger species such as the rat and rabbit. The preferred predatory animal would be the farm cat. The predator animal must be aggressively predatory but also able to be handled by the licensee. The farm cat is ideal being semi-feral and able to survive on its hunting ability. The normal, well-fed house cat is unlikely to be sufficiently interested in the prey, if it recognises it as prey at all.

Protocol – three types of study will be carried out:

A floor pen with sides 2 m wide, 3 m long and 1.5 m high, made of galvanised metal, will be used. It will be divided along the centre of its length by a 10 cm high wire mesh (1 cm × 1 cm squares) on top of which will be a transparent sheet of plastic which reaches to the top of the pen enabling clear vision between the two areas but no physical access (Figure 5.1). Mice will be confined to the front pen and the predator to the back pen. Food and water will be provided in trays placed next to the dividing mesh to ensure that mice have to approach the area of the predator. Sawdust bedding, but no other cover, will be provided.

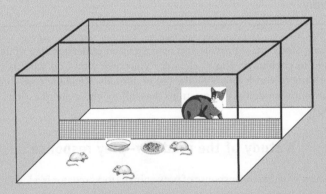

Figure 5.1   Predator–prey response enclosure

(1) **Normal control behaviour of mice in absence of predator odour:** First, the normal behaviour of each group of mice when in the front pen, with no predator in the back pen, will be determined as a control. Each group will contain 25 mice that will be held in the pen for 48 hours. Behaviour, in particular time spent close to the partition, will be observed using an infrared camera and video recorder. Faeces samples will be collected for cortical assays to measure stress levels.

(2) **Behaviour in presence of predator odour:** All mice will be removed from the front pen and the predator placed in the back pen for 24 hours. Urine and faeces produced by the predator will be left in the pen when the predator is removed. Each group of mice, in turn, will then be held in the front pen for 48 hours and behaviour monitored as in (1). Faeces samples will be collected as above.

(3) **Behaviour in presence of predator:** The predator will be kept in the back pen and each group of mice in turn again held for 48 hours in the front pen. Behaviour will be monitored as in (1). The predator will be fed in the pen and offered surplus mice killed by a Schedule 1 method. It is a necessary part of the learning experience of each group of mice that they see the predator consuming mouse carcasses. Faeces samples will be collected as above.

Now consider, with some of your fellow students, the following questions:

(1) Is this study actually **necessary** or at least important for some well-defined and recognised purpose?

(2) Is the specified protocol acceptable or are there flaws that require it to be improved?
Considerations:

   (a) Is it scientifically necessary and ethically justifiable to locate the prey's food in such close proximity to the predator?

   (b) Would it be sufficient just to leave the predator's faeces and urine at a discreet distance?

   (c) Should the mice be given nesting material as well as sawdust?

   (d) Is it strictly necessary for each group of mice to see the predator consuming mouse carcasses?

   (e) Is there sufficient provision for removing the scent of mouse groups so that trials with successive cohorts are independent (mistakes on this point might require all experiments to be repeated)?

   (f) Is the choice of mice as the prey species adequately justified? Is their cheapness a relevant consideration?

   (g) Is the number of mice to be used adequately justified?

   (h) Is the choice of predator adequately justified?

   (i) Is it sufficiently clear what would happen to the mice at the end of the study? If not, what should be specified?

5.8.2 *Tutor notes*

Place the students in groups of four or five to consider the questions posed before bringing the groups together to discuss these collectively. Possible answers to the questions are as follows:

(1) Is this study actually **necessary** or at least important for some well-defined and recognised purpose?

- The Animal (Scientific Procedures) Act allows not only for medically related studies but also for studies of animal behaviour.

- An understanding of learned behaviours is important if we are to understand human development.

  Can it be argued that the study will cause needless suffering to mice, and thus should not be approved?

- The study may cause stress/anxiety but not more than that seen in the wild, and the animals will not be subjected to physical pain.

(2) Is the specified protocol acceptable or are there flaws that require it to be improved? Considerations:

(a) Is it scientifically necessary and ethically justifiable to locate the prey's food in such close proximity to the predator?

- No to both.

(b) Would it be sufficient just to leave the predator's faeces and urine at a discreet distance?

- Depends on air movement and the size of the chamber.

(c) Should the mice be given nesting material as well as sawdust?

- Yes, or the study will involve an unnatural environment and additional, unnecessary stresses.

(d) Is it strictly necessary for each group of mice to see the predator consuming mouse carcasses?

- Visual sensory cues are as important as other sensory cues such as smell.

(e) Is there sufficient provision for removing the scent of mouse groups so that trials with successive cohorts are independent (mistakes on this point might require all experiments to be repeated)?

- No; removal of bedding and thorough cleaning of each pen should be carried out after each group of mice is removed.

(f) Is the choice of mice as the prey species adequately justified? Is their cheapness a relevant consideration?

- Cost is not a consideration here; the Act requires that the lowest sentient species (less able to perceive or sense things; for example, a rat is viewed as less sentient than a monkey) and most appropriate species should be used. In this instance mice are an appropriate choice as they are the normal prey of the predator and, along with rats, are considered to have the lowest level of sentience of all animals used in research.

(g) Is the number of mice to be used adequately justified?

- Requires power calculations.

(h) Is the choice of predator adequately justified?

- Yes, natural predator.

(i) Is it sufficiently clear what would happen to the mice at the end of the study? If not, what should be specified?

- Recognised method of humane killing e.g. Schedule 1 in the UK.

5.9 Harm/benefit analysis

The purpose of a harm/benefit analysis of a proposed programme of research work (undertaken by the Institutional Animal Care and Use Committees (IACUCs) in the US, or the local Ethical Review Process and subsequently the Home Office Inspectorate in the UK), is to ensure that potential harm to the animals being used does not outweigh the benefits of the research to science or society. These committees undertake a robust analysis of a project to ensure that the experimental methodology and design is appropriate and that all potential harm to the protected animals over their entire lifetime has been identified and avoided where possible or, if it can't be avoided, that mechanisms are put in place to alleviate the harm. They also look for evidence that: the 3Rs have been fully implemented; there is no alternative to the use of animals in the research; the choice of species is justified and appropriate; staff are suitably trained; there are suitable facilities available to undertake the work; and that the project will comply with an institution's policies and procedures (e.g. on tail tipping (cutting off the tip of the tail, extracting and analysing DNA from it to determine the genetic phenotype of the animal), use of anaesthetics, inspection regimes etc.).

5.9.1 Tutor notes

You may wish to get the students to undertake a harm/benefit analysis of the spinal cord injury case study (Section 5.6.1). Following consideration of the potential harm the project may cause to animals (see the harm/benefit analysis paragraph above for points that might be considered), they should weigh this against the potential benefits, not only to spinal cord injury patients but also to those around them (parents, brothers/sisters, carers). They should also consider any religious or societal perspectives on the study, including controversy that may arise through the use of stem cells.

5.10 The arguments for and against animal experimentation

The use of animals in scientific and medical research is controversial, arousing strong emotions in individuals either fundamentally opposed to, or strongly supportive of, their use. It is essential that research scientists who use either intact animals or animal tissues in their work should have knowledge of the common arguments for and against the use of animals in research. Should the need arise, this will allow them to defend their own use of animals and, more importantly, it will mean they have considered their own opinions on the subject and will know how far they are prepared to go in future experiments with animals. For example, they may decide that there are particular techniques and approaches involving certain species (dogs, monkeys?) that they personally would not be prepared to use.

5.10.1 Case study

The aims of this case study are to:

- explore the case for, and the case against, the use of animals in scientific and medical research;
- increase your awareness of the ethical issues and legislative requirements surrounding the use of animals in research.

Your tutor will divide the class into four groups, giving each group a particular 'cause' (scientist; patients and clinicians; animal welfare and the 3Rs; antivivisectionists). Using the links (URLs) below as a starting point, your group should prepare a seven-minute presentation to promote your 'cause'/point of view. The presentation should not just be a list of facts but should be a reasoned, logical argument which seeks to persuade the audience to a particular point of view (your 'cause'). It should not simply consist of a list of groups that support your cause, or a list of their aims or objectives. Instead, it should comprise an introduction, the main body of the talk, which provides evidence to support your cause, and a conclusion. You can use emotional arguments but these should be supported by scientific facts. Even if you don't agree with your allocated viewpoint, for the purposes of this presentation you should adopt the position of a committed believer of your cause.

Following these presentations, your tutor will facilitate a debate or discussion around the issues raised.

Antivivisectionists campaigning against the use of animals in scientific research:

- British Union for the Abolition of Vivisection www.buav.org/
- Animal Liberation Front www.animalliberationfront.com/
- SPEAK www.speakcampaigns.org/index.php/
- National Anti-Vivisection Alliance www.antivivisection.info/
- People for the Ethical Treatment of Animals www.peta.org.uk/
- American Anti-Vivisection Society www.aavs.org/index.php.

Individuals promoting animal welfare and the 3Rs (refinement, reduction and replacement of animal use in research):

- Fund for the replacement of animals in medical experiments www.frame.org.uk
- RSPCA www.rspca.org.uk

- National Centre for the 3Rs www.nc3rs.org.uk/

- Universities Federation for Animal Welfare www.ufaw.org.uk/.

Scientists explaining the need to use animals in scientific and medical research:

- Understanding Animal Research www.understandinganimalresearch.org.uk/

- Association of Medical Research Charities www.amrc.org.uk

- Foundation for Biomedical Research www.fbresearch.org.

Patients and clinicians supporting the need for animal studies:

- Pro-Test www.pro-test.org.uk/

- Coalition for medical progress www.medicalprogress.org/

- Americans for Medical Progress www.amprogress.org/content/home.

Additional links can be found at the Understanding Animal Research (www.understandinganimalresearch.org.uk/) and the Home Office (http://scienceandresearch.homeoffice.gov.uk/animalresearch/reference/) websites.

5.10.2 Tutor notes

This seminar can be used to increase understanding and ethical awareness both of individuals who may use intact animals and also those who may use animal tissues in their research or studies.

Students should be given sufficient time before the seminar to prepare their groups presentations. In the seminar itself, there are two options:

(1) The presentations can be used to provide background knowledge prior to a discussion of the issues raised which could include: why use animals rather than alternatives?; the ethics of such studies; animal welfare and the 3Rs; animal welfare legislation in the UK or elsewhere; the antivivisection movement and legitimate protests. For further details, see 'Can the use of animals in scientific research be justified' in the open educational resource *Animal welfare, ethics and the 3Rs: training materials and resources* (Lewis, 2009).

(2) Alternatively, students can remain in their allocated groups and, after the presentations, each group should be given 15 minutes to defend their 'cause' in response to questions from opposing groups. You may wish to facilitate this process by requiring each group to pre-circulate a summary of the main points of their presentation to opposing groups 48 hours before the seminar so that the latter have the opportunity to prepare questions for the debate. Tutors may be required to facilitate this debate (by being provocative), but it should be as student-led as possible.

5.11   How you can achieve your potential in *in vivo* work

- Understand the place of *in vivo* studies in modern biomedical research.

- Ensure you are familiar with the laws and regulations governing the use of animals in research.

- Understand the arguments for and against the use of animals; be familiar with relevant ethical issues and review procedures.

- Understand the principles of the 3Rs and explore alternatives to the use of animals.

- Enrol on *in vivo* skills training courses and workshops (see Section 5.13, Additional resources).

5.12   References

Attarwala, H. (2010) TGN1412: from discovery to disaster. *Journal of Young Pharmacists: JYP* **2**, 332–336.

Botting, J. (2002) The history of thalidomide. *Drug News & Perspectives* **15**, 604–611.

British Pharmacological Society and the Physiological Society (2006) Tackling the need to teach integrative pharmacology and physiology: problems and ways forward. *Trends in Pharmacological Sciences* **27**, 130–133.

Association of the British Pharmaceutical Industry and the Biosciences Federation (2007) *In Vivo Sciences in the UK: Sustaining the Supply of Skills in the 21st Century*. London: ABPI/Biosciences Federation. Available at: www.abpi.org.uk/our-work/library/industry/Pages/in-vivo-report.aspx. Accessed April 2011.

Centers for Disease Control and Prevention (2011) U. S. Public Health Service Syphilis Study at Tuskegee (www.cdc.gov/tuskegee/index.html). Accessed April 2011.

Council of Europe (2007) Convention for the Protection Of Human Rights and Dignity of the Human Being with Regard to the Application of Biology and Medicine: Convention on Human Rights and Biomedicine CETS No.: 164 (http://conventions.coe.int/Treaty/Commun/QueVoulezVous.asp?NT=164&CL=ENG). Accessed April 2011.

ECVAM (2007) European Centre for the Validation of Alternative Methods Scientific Advisory Committee Statement on the Validity of *In-Vitro* Tests for Skin Irritation. (http://ecvam.jrc.it/ft_doc/ESAC26_statement_SkinIrritation_20070525_C-SEC.pdf). Accessed April 2011.

European Union (1986) Council Directive 86/609/EEC of 24 November 1986 on the Approximation of Laws, Regulations and Administrative Provisions of the Member States Regarding the Protection of Animals Used for Experimental and Other Scientific Purposes. (http://eur-lex.europa.eu/LexUriServ/LexUriServ.do?uri=CELEX:31986L0609:EN:HTML). Accessed April 2011.

European Union (2010) Directive 2010/63/EU of the European Parliament and of the Council of 22 September 2010 on the Protection of Animals Used for Scientific Purposes (http://eur-lex.europa.eu/LexUriServ/LexUriServ.do?uri=OJ:L:2010:276:0033:0079:EN:PDF). Accessed April 2011.

Frye, J., Ananthanarayanan, R. and Modha, D.S. (2007) Towards real-time, mouse-scale cortical simulations. *IBM Research Report RJ10404 (A0702-001)*, IBM Research Division, San Jose, CA. Available at: www.almaden.ibm.com/cs/people/dmodha/rj10404.pdf. Accessed April 2011.

HM Government (1986) Animal (Scientific Procedures) Act 1986 (http://tna.europarchive.org/20100413151426/http://www.archive.official-documents.co.uk/document/hoc/321/321-xa.htm). Accessed April 2011.

Home Office (1998) The Ethical Review Process (http://webarchive.nationalarchives.gov.uk/ + /http://www.homeoffice.gov.uk/docs/erpstatement.html). Accessed April 2011.

Lewis, D.I., (2009) Animal welfare, ethics and the 3Rs: training materials and resources. Interdisciplinary Ethics Applied Centre of Excellence in Learning and Learning (IDEA CETL), University of Leeds. Available at: www.lasa.co.uk/3Rs%20resource%20e-version%20Dec09s.pdf. Accessed April 2011.

Madsen, A.N., Hansen, G. Paulsen, S.J., *et al.* (2010) Long-term characterization of the diet-induced obese and diet-resistant rat model: a polygenetic rat model mimicking the human obesity syndrome. *The Journal of Endocrinology* **206**, 287–296.

Russell, W.M.S. and Burch, R.L. (1959) *The Principles of Humane Experimental Technique*. London: Methuen.

US Department of Agriculture (1966) Animal Welfare Act (http://awic.nal.usda.gov/nal_display/index.php?
   info_center=3&tax_level=3&tax_subject=182&topic_id=1118&level3_id=6735&level4_id=0&level5
   _id=0&placement_default=0). Accessed April 2011.
US Department of Agriculture (2009) A Quick Reference of the Responsibilities & Functions of the Institutional
   Animal Care & Use Committee (IACUC) Under the Animal Welfare Act (www.nal.usda.gov/awic/pubs/
   Legislat/awabrief.shtml#Q15). Accessed April 2011.
US Department of Health and Human Services (2011) Information on the 1946–1948 United States Public
   Health Service STD Inoculation Study (www.hhs.gov/1946inoculationstudy/). Accessed April 2011.

5.13   Additional resources (accessed April 2011)

5.13.1   Alternatives to the use of animals and the 3Rs

ALTBIB - Resources on Alternatives to the Use of Live Vertebrates in Biomedical Research and Testing. National
   Library of Medicine. http://toxnet.nlm.nih.gov/altbib.html.
ALTWEB. John Hopkins Bloomberg School of Public Health gateway to educational and research resources on
   alternatives to animal testing. Includes a searchable database. http://altweb.jhsph.edu/.
European Centre for the Validation of Alternative Methods (ECVAM). Provides factual and evaluated information
   on advanced non-animal methods for toxicology assessments; offers full method descriptions, including
   development and validation status. (Includes INVITTOX test protocols.) http://ecvam.jrc.it/.
Fund for the Replacement of Animals in Medical Experiments (FRAME). Materials and resources on alternatives to
   the use of animals and the promotion of the 3Rs. www.frame.org.uk/index.php.
National Centre for the Replacement, Refinement and Reduction of Animals in Research (NC3Rs). Resources for
   implementation/good practice in all aspects of the 3Rs. www.nc3rs.org.uk.
The Norwegian Reference Centre for Laboratory Animal Science & Alternatives. Information and resources on
   laboratory animal science and alternatives to the use of animals in research, teaching and school dissection
   classes. http://oslovet.veths.no/dokument.aspx?dokument=80. See http://film.oslovet.veths.no for video
   clips.

5.13.2   Animal welfare legislation

5.13.2.1   Animal Welfare Act 1966 (and amendments) (US)

Animal Welfare Act. USDA National Agricultural Library Animal Welfare Center. http://awic.nal.usda.gov/
   nal_display/index.php?info_center=3&tax_level=3&tax_subject=182&topic_id=1118&level3_id=6735
   &level4_id=0&level5_id=0&placement_default=0.
Regulations: Title 9—Animals and Animal Products. USDA Animal and Plant Health Inspection Service. www.
   aphis.usda.gov/animal_welfare/downloads/awr/awr.pdf.
Guidance: Animal Welfare Act Information. USDA Animal and Plant Health Inspection Service. www.aphis.usda.
   gov/animal_welfare/awa_info.shtml.
Overview: The Animal Welfare Act: An Overview. USDA Animal and Plant Health Inspection Service. www.aphis.
   usda.gov/publications/animal_welfare/content/printable_version/animal_ welfare4-06.pdf.

5.13.2.2   Animal (Scientific Procedures) Act 1986 (UK)

Animals (Scientific Procedures) Act 1986. The National Archive. http://tna.europarchive.org/
   20100413151426/http://www.archive.official-documents.co.uk/document/hoc/321/321-xa.htm.
Guidance: Guidance on the Operation of the Animals (Scientific Procedures) Act 1986. The National Archive.
   http://tna.europarchive.org/20100413151426/http://www.archive.official-documents.co.uk/document/
   hoc/321/321.htm.
Animal (Scientific Procedures) Division: Research and Testing Using Animals. The Home Office. www.
   homeoffice.gov.uk/science-research/animal-research/.

5.13.2.3   European Union

ETS 123: European Convention for the Protection of Vertebrate Animals Used for Experimental and Other Scientific Purposes. Strasbourg, 18.III.1986. http://conventions.coe.int/Treaty/en/Treaties/html/123.htm.

EU Directive 2010/63/EU: Directive 2010/63/EU of the European Parliament and of the Council Of 22 September 2010 on the Protection of Animals Used for Scientific Purposes. http://eur-lex.europa.eu/LexUriServ/LexUriServ.do?uri=OJ:L:2010:276:0033:0079:EN:PDF.

EU Directive 86/609/EEC: Council Directive 86/609/EEC of 24 November 1986 on the Approximation of Laws, Regulations and Administrative Provisions of the Member States Regarding the Protection of Animals Used for Experimental and Other Scientific Purposes. http://eur-lex.europa.eu/LexUriServ/LexUriServ.do?uri=CELEX:31986L0609:EN:HTML.

Guidance: Laboratory Animals. European Commission Environment. http://ec.europa.eu/environment/chemicals/lab_animals/home_en.htm.

5.13.3   Animal welfare organisations

Royal Society for the Prevention of Cruelty to Animals (RSPCA) Science Group. Resources for improving the welfare of wildlife, laboratory, farm and companion animals. www.rspca.org.uk/sciencegroup.

Universities Federation for Animal Welfare (UFAW). Resources for improving the welfare of animals as pets, in zoos, laboratories, on farms and in the wild. www.ufaw.org.uk/index.php.

Animal Welfare Institute (AWI). Organisation which seeks to reduce the suffering of laboratory, farm and captive animals. www.awionline.org/ht/d/sp/i/214/pid/214.

Dr Hadwen Trust for Humane Research. Medical research charity that funds and promotes exclusively non-animal techniques to replace animal experiments. http://drhadwentrust.org/.

5.13.4   Training and education

American Association for Laboratory Animal Science. Training resources and guidance on the use of animals in research. www.aalas.org.

Association for Assessment and Accreditation of Laboratory Animal Care (AAALAC) International. Resources and information on the provision of training in research animal care and welfare. www.aaalac.org/.

Federation of European Laboratory Animal Science Associations (FELASA). Includes policy documents, training resources and guidance on animal welfare and the 3Rs. www.felasa.eu/.

Laboratory Animal Science Association (LASA, UK). Includes resources, reports, guidance for implementation of best practice in 3Rs and Animals (Scientific Procedures) Act. www.lasa.co.uk/index.html.

5.13.5   Science and society

Americans for Medical Progress (AMP). Resources to promote understanding and acceptance of the need for humane animal research in the US. Factsheets, DVDs and other resources. www.amprogress.org.

Foundation for Biomedical Research (FBR). Resources for use in public education and information on the use of animals in research. www.fbresearch.org.

Understanding Animal Research (UAR). Resources to promote understanding and acceptance of the need for humane animal research in the UK. Factsheets, videos and images of the use of animals in research. Speakers in schools presentations and training. www.understandinganimalresearch.org.uk/homepage.

6 Research projects

Martin Luck

Prologue

A three-stage dialogue between Student and Supervisor, about a final-year research project in animal science.

Stage 1

STUDENT: 'Hello. I'd like to do a project on African mammals. I'm really interested in elephant.'

SUPERVISOR: 'That sounds good. What area are you particularly interested in?'

STUDENT: 'Well, something to do with reproduction. I had a holiday job at the zoo and I heard they don't breed in captivity.'

SUPERVISOR: 'I've heard that too. Do you know why? I expect there is some literature on this. You could try to find out what the main problems are and suggest ways around it.'

STUDENT: 'Wow! That would be exciting. What books should I look at?'

SUPERVISOR: 'Why not look in the *Zoo Yearbook* for the latest data on captive breeding programmes, and at the same time search the literature and see what people are studying. I'll show you how to start. You could also talk to your friend at the zoo and see what she knows.'

STUDENT: 'These papers I've found suggest it's to do with behaviour, but they also think there might be problems with mating, or nutrition, or social grouping or age. Some think it's the males who are infertile, others the females. The zookeeper thinks it's too cold for them here. It's all so confusing.'

SUPERVISOR: 'OK. Well let's start by trying to understand how the elephant reproductive system works . . .'

Effective Learning in the Life Sciences: How Students Can Achieve Their Full Potential, First Edition. Edited by David J. Adams.
© 2011 John Wiley & Sons, Ltd. Published 2011 by John Wiley & Sons, Ltd.

Stage 2

STUDENT: 'I've found lots of info on the anatomy of the elephant reproductive tract, and on the female cycle and pregnancy. It's really complicated: a bit different from other mammals. The research papers are also saying that they are prone to ovarian disease, especially as they get older.'

SUPERVISOR: 'Are most of the females in zoos old or young? How long have they been in captivity? Perhaps the zoos are trying to breed the wrong individuals.'

STUDENT: 'Yes, the zoo data seems to suggest that. But I also found this paper suggesting that the incidence of poor fertility is higher in zoos than in the wild, for animals of the same age.'

SUPERVISOR: 'What do you think that could be due to?'

STUDENT: 'Some papers are suggesting that the social hierarchy influences mating frequency and the chance of a successful mating.'

SUPERVISOR: 'Well, that's common in other animals. Hormones from the brain affect ovary function, and some of these hormones are affected by pecking order. I wonder if that could be the case in elephant.'

STUDENT: 'I'll see what I can find. By the way, I don't think I'll have time to investigate male problems, or the effects of climate. There is just too much to read.'

SUPERVISOR: 'That's fine. Why not just focus on the link between hormones and female reproductive success. It will give you plenty to write about. You've found lots of papers on that already. You can make a critical appraisal of the published info, decide if it's adequate, and suggest what more needs to be done.'

Stage 3

STUDENT: 'Hello. I've had a really productive few weeks, writing up a description of the female tract and looking at all that literature on brain hormones and reproduction.'

SUPERVISOR: 'What have you decided?'

STUDENT: 'Well, the anatomy is described in great detail and many diseases are reported, but not much is known about elephant hormones. I expect it's hard to get samples over the cycle, or from lots of different individuals. The info on other mammals is much better. Can I extrapolate it to elephant?'

SUPERVISOR: 'Not safely: as you have discovered, the reproductive system varies a lot between species.'

STUDENT: 'Yes, I'll have to acknowledge that and make do with what there is. Oh, by the way, I've had another idea. We had a lecture last week on leptin, the nutrition and body fat hormone. We were told that it also affects reproduction. I can't find anything on leptin in elephant, but I wonder if zoo elephants have the wrong body weight to be fertile?'

SUPERVISOR: 'That's a really interesting idea. There probably isn't any direct research on it in elephant but you could set up a hypothesis and suggest a way to test it. Use this as the material for your Discussion.'

STUDENT: 'I'm excited that I thought of that myself. No one seems to have done so before.'

SUPERVISOR: 'I agree. It's your idea. You've done lots of solid background research, so now you can put it forward with confidence. This has been a valuable project.'

Coda

This dialogue summarises a real conversation. The project was eventually written up as a research report and placed in the public domain (www.nottingham.ac.uk/burn/Noon.pdf). The student subsequently studied for a Masters degree in wild animal biology and used her ideas as the basis for a laboratory-based research project.

6.1 Introduction

If you are in the final stages of your degree course, it is likely that you are about to undertake a research project. This is an exciting challenge but you will have many questions. It is a completely different way of studying from the formality of lectures and practical classes, and therefore unfamiliar personal territory. You will want to have a clear understanding of what you are doing and its purpose, so that you can make the most of the opportunity.

This chapter tries to explain why research is considered such an important part of the degree programme, and tackles some of the issues you might be concerned about. It also suggests some questions you might want to ask about your project before you start, to make sure it's right for you and to ensure that you get the most out of it. Your project will be unique; it will have research goals but it must also be a satisfactory educational experience. You and your supervisor will need to have a clear picture of what you are trying to achieve and work together to make it a success.

The other elements of your degree are likely to have been based around directed learning of one kind or another. Although essential for subject knowledge and sound understanding, that kind of study is unlikely by itself to make you feel that you have fully engaged with the subject and that it belongs to you. You might have studied biochemistry or ecology, but to call yourself a biochemist or an ecologist you need to investigate the subject yourself. Thus your project is a chance to get deeply involved and make an original contribution to a field of research. If all goes well, you may find that you have taken ownership of your subject area, in that you begin to feel personally involved with it as a scientific discipline.

Most students find their research project to be a positive and enjoyable experience. For many, it is the defining element of the study programme. It is as much a personal journey as an educational and scientific challenge. Many students use the experience to help them decide actively to pursue, or not to pursue, a career involving research. Either way, the whole experience is likely to be a defining one for your degree, your career and your future (Jenkins *et al.*, 2003; Luck, 2009).

6.2 Research project – role and purpose

It is common for UK undergraduate students, particularly in the sciences, to undertake research. Most UK degree programmes incorporate an investigative project, often in the final year. The Quality Assurance Agency (QAA) Honours Degree Benchmark Statements (www.qaa.ac.uk/academicinfrastructure/benchmark/honours/default.asp) require graduates to have experienced research, and the Society of Biology is likely to require research to be a component of its accredited degrees (www.societyofbiology.org/education). This contrasts with much of the USA, where research is frequently an optional extra (Taraban and Blanton, 2008). It is not hard to see why academia attaches such importance to student research: it offers you a clear path towards intellectual independence, a deep understanding of your subject and the development of a wide range of personal, intellectual and transferable skills.

Table 6.1   The benefits of doing a research project: what's in it for you?

Ownership	Contribute to your chosen discipline, be part of it and make it your own.
Real science	Make discoveries; learn at first hand how science works; appreciate how new understanding is gained; stand at the edge of confident knowledge.
Skill development	Develop a range of skills to do with: • process (working as a researcher) • management (planning, organising, doing, exploiting) • presentation (telling others about it, writing and presenting) • other personal attributes (extending your abilities, achieving more, gaining confidence, dealing with criticism, working with others).
Enjoyment	Gain independence, be motivated and have fun.

Table 6.1 summarises some of the benefits you might expect to gain from carrying out a research project. Many of these things, especially the skills, have been discussed at length elsewhere (Luck, 1999, 2009; Ryder, 2004). Their exact nature will depend on your subject area, the sort of academic environment in which you are studying and the precise nature of the project topic. Prepare to be moved out of your comfort zone: the project will stretch your abilities and you will achieve things you never dreamed of.

First and foremost, your research project is part of your degree and needs to have sound educational characteristics. For example, whilst you will learn much from it, it needs to fit with your course and it must be assessed fairly and objectively. All other considerations have to be of secondary importance, including the utility of the research itself. This may sound strange, given the enthusiasm you probably feel about finally getting down to proper science. But, as we shall see, the aim is to give you the *experience* of doing research, irrespective of the actual outcome of the project. You should expect your supervisor to be aware of this and to have considered it when designing your project or moderating your research plans.

6.3   Applying the Scientific Method

By doing research yourself, you will be exposed to the uncertainties, excitements and vicissitudes of scientific progress and will come to know exactly where the limits of confident knowledge are located. It will give you personal experience of how new knowledge is acquired and how deep understanding is achieved.

Your project should bring you face to face with the **Scientific Method** (Box 6.1). The Scientific Method is interpreted by practising scientists and academics in diverse ways. For example, many supervisors require students to make clear statements of hypotheses and to test them formally (perhaps in a more explicit manner than the supervisor might normally do in their own day-to-day research). If you are asked to do this it will force you to confront, with clarity and precision, the question your investigation is asking and the true value of the results you obtain. Other supervisors feel that the Scientific Method is best experienced more by *doing* than by abstract thinking. Whatever your supervisor's approach, it is essential that your project should give you a real scientific experience: it should achieve a balance between subjective speculation and uncertainty on the one hand, and objective progression on the other.

Box 6.1   The Scientific Method

During the last 500 years, scientific research has been an extraordinarily successful human endeavour as scientists have devised and modified strategies for investigation and the interpretation of results. An approach that embraces the best of these strategies is known as the Scientific Method (Luck and Wagstaff, 2003). A key element of the approach is 'hypothesis testing'.

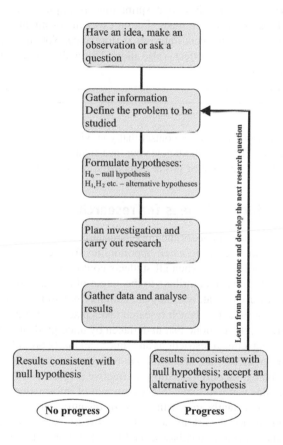

Hypotheses predict the possible effects of an independent variable (a 'treatment') on a dependent variable (something being measured or monitored). A series of hypotheses can be formulated to cover all possible outcomes. The null hypothesis (H_0) postulates that there will be no statistically significant difference between the results obtained in the presence or absence of the treatment. In the first instance, experiments are designed to establish whether the null hypothesis can be rejected. If H_0 cannot be rejected, the current situation persists and no progress is made. However, if the results indicate that H_0 can be rejected, one of the alternative hypotheses (H_1, H_2 etc.) can be accepted and there is an advance in our knowledge and understanding.

Example

Question: Does multivitamin supplementation enhance athletic performance?

H_0: Multivitamin supplementation has no effect on performance in a sprint exercise.

H_1: Multivitamin supplementation improves performance in a sprint exercise.

H_2: Multivitamin supplementation reduces performance in a sprint exercise.

The results of the experiment show that sprint time is improved after multivitamin supplementation. Progress has been made. Future experiments might test whether each individual vitamin in the supplement improves performance. Note that the possibility of supplements having a detrimental effect is also allowed for within the design and expressed amongst the hypotheses.

It is also worth bearing in mind from the outset that you will have to write-up the whole project experience in a dissertation. Thinking about how your investigation will appear when exposed to the cold light of assessment, or as an oral or poster demonstration, should help focus your thoughts!

6.4   Types of project and ideas for research

Pressure from increasing student numbers and limited resources means that departments have to be inventive and resourceful in offering research opportunities for their students. You may perhaps imagine that there was a golden age when all students could do a well-resourced, experimental project in their supervisor's lab and contribute to the output of an established research group. It is doubtful if things were ever like that, and they are certainly not so now. Many students study in departments where little or no active bench research takes place, and staff often face inadequate funding for their own research, let alone that undertaken by undergraduates. Issues such as health and safety, animal licensing, ethical restrictions and time availability provide further constraints on what is possible. However, as we shall see, while lab-based research can be a great experience for students, it is far from being the only scenario for effective project work.

From your recent studies you will know that the amount of information generated in the many bioscience disciplines is constantly increasing both in quantity and diversity. New knowledge continues to come from empirical work in the lab but it also arises from surveys and field studies, through interrogation and analysis of databases, and by exploitation of literature and other material already in the public domain. The biosciences also have a strong presence in education, public sector enterprise and industry. All of these environments provide ideal opportunities for student research.

The diversity of research projects currently offered by UK bioscience departments is illustrated in Table 6.2. Further case studies can be found in Luck (2009) and related resources (www. bioscience.heacademy.ac.uk/resources/guides/studentrescs.aspx). What is remarkable is the imaginative way in which staff use their experience, contacts and resources to provide flexible, exciting and challenging research programmes. Each represents the reality of contemporary bioscience and can give you an experience as valuable as that to be obtained wearing white coat and goggles. By carefully structuring the research environment, and having defined learning objectives, you and your supervisor can ensure that you engage realistically and effectively with the process of scientific enquiry.

Table 6.2 Types of undergraduate research project, with snapshots based on real examples and case studies

Type of project	Example/case study
Experimental lab work	(1) Investigation of phosphatase Pp2A as regulator of anaphase in a human cell line: research involved manipulation of DNA and time-lapse microscopy of cell division; concluded that Pp2A activity is needed for chromatin separation. (2) Comparison of the effects of tomato varieties with different flavonoid content on growth of human colon cancer cells: cells were exposed to tomato extracts and proliferation measured; all extracts inhibited cell division, but according to total polyphenol content, rather than the identities of individual flavonoids.
Bioinformatics and database analysis	(1) Analysis of social relationships among zoo elephants using Social Network Analysis software: behavioural data from video recordings revealed complex dominance hierarchies and small cliques associated with specific behaviours; results may be valuable in animal management. (2) Analysis of co-authorship amongst academic staff to measure the interdisciplinarity of research: used cluster analysis of data from Web of Science (http://wokinfo.com/products_tools/multidisciplinary/webofscience/); concluded that academics could be realistically grouped by research focus and that the extent of collaboration was greater amongst chemists than amongst biologists.
Fieldwork and expeditions	(1) Assessment of orchid populations in different regions of a post-industrial country park with SSSI status: results showed regional variations in orchid morphology related to soil pH and scrub management practices. (2) Effects of power-station outflows on invertebrate populations in river water: study involved water sampling and insect larvae counts upstream and downstream of a power station; a key finding was that larvae counts were related to water temperature.
Opinion surveys	(1) Assessment of attitudes of members of the public to stem cell technology: questionnaire used to obtain views on specific stem cell applications in medicine, together with information on respondents' level of scientific education; analysis showed greater acceptance amongst science-educated respondents. (2) Analysis of consumer attitudes to welfare of farm animals: a survey of consumers measured levels of concern about pig welfare and knowledge of pork production systems; concluded that demand for high-welfare pork would be increased by greater knowledge of production methods.
Commercial collaboration and product development	(1) Handheld biosensor technology as anti-terrorism devices: student compared available technologies and performed a computer analysis of number of organisms detected, detection time, price and unit size. (2) Use of thin layer films of organic materials for topical drug delivery: working with pharma company, student investigated types of drugs that could be delivered, patients who might benefit and ways of altering drug dosage by film overlaying.

(*continued*)

Table 6.2 (*Continued*)

Type of project	Example/case study
Literature review	(1) A summary of literature on reproductive strategies of insectivorous bats: categorised reproductive systems in relation to habitat and latitude; proposed links between evolution of diapause and light, rainfall and diet. (2) A meta-analysis of studies on the nutritional value of vegan and vegetarian diets: collected and summarised published reports on vitamin content of these diets; assessed likely implications of non-animal diets for human health and the need for vitamin supplementation.
Public and educational communication	(1) Development of public information resources on importance of vitamin D: based on literature indicating possible vitamin D deficiency amongst some ethnic communities with reduced skin exposure to light; designed leaflets for potential distribution in GP surgeries and community centres. (2) Design and delivery of a lesson for Year 10 school pupils on the dairy industry and milk production: surveys of pupil knowledge of milk production before and after the lesson ascertained its effectiveness.
Development of e-learning resources	(1) Creation of a knowledge database on growth factors, as a reference resource for students studying second-level endocrinology: undertaken by three students, each covering a different class of growth factors. (2) Creation of interactive learning resource for nursing students, describing the importance of differential white cell counts in clinical assessments: project used an Analysis, Design, Development, Implementation and Evaluation (ADDIE) heuristic design model; effectiveness was assessed by pre- and post-use knowledge questionnaires.

6.5 Characteristics of good research projects

So, how can you be sure that your project will offer all of the benefits you might hope for, bearing in mind that you can choose only once from what may be several attractive options? The sheer diversity of topics and methods can be confusing and it may be hard to evaluate the various possibilities. Each of the projects on offer should allow you to acquire knowledge and develop new skills, but this alone does not justify or validate any particular investigation. You may also be basing your choice on personal criteria such as interests, aptitude and career aspirations, and you might not yet have a clear picture of the actual research work each project will involve.

With this in mind, it is helpful to have some criteria to help you decide whether a project is right for you. There are four key questions you might ask, and I now consider these and provide some indication of the sort of answers you might look for. The questions are summarised in Table 6.3, which you can use as a 'checklist' as you consider the projects on offer at your university. Your supervisor faces similar challenges when thinking of project topics and designing suitable investigations. Additional criteria, from the supervisor's perspective, are suggested in the notes for tutors.

Table 6.3 Project checklist

Will this project suit me? Do I know what I'm doing?	
Is it a proper research experience?	• Is this an opportunity for real research? • Is the research interesting, exciting and relevant to my interests? • Will I have the opportunity to be creative? • Does the investigative approach clearly embrace the Scientific Method?
Does it have a clear structure?	• Is there a clear overall target/envisaged outcome? • Are objectives defined clearly? • Is there a well-structured investigative programme?
Are risks identified and responsibilities defined?	• Have intellectual and practical risks been anticipated and allowed for? • Are all of the resources required for this project readily available? • Will I be given training in appropriate techniques? • Is the academic supervisor clearly identified? • Will others e.g. postdoctoral fellows, postgraduate students, be involved in the project? • Am I entirely clear about what will be expected of me during the time involved?
Will I be successful?	• Is there a real opportunity to succeed? • Are there opportunities to excel, and will exceptional performance be rewarded?

6.5.1 Is it a proper research experience?

Undergraduate research should be real: it should be a genuine attempt to discover something new, develop understanding or create a novel product. It should involve the application of the Scientific Method and offer the possibility of reaching a clear conclusion. Although your project may have been designed with a student in mind, it should not be contrived for the sake of study, nor should it be a pointless procedural exercise: it needs to be genuinely creative. As you moved from science at school to university studies, you will have seen that the purpose of research is not to demonstrate the known but to pursue the unknown. An experiment is only worthwhile if the outcome is uncertain. Equally, a published dataset or body of literature has utility only when it is interpreted or exploited, and no single interpretation is objectively correct. The test of authenticity for a student project is therefore to ask 'Is this project scientifically creative? Does it present the possibility of novelty or discovery?' Asking this early in the investigation can be a good way of starting to think deeply about the research you are about to undertake.

A further aspect of authenticity is the extent to which you are able to engage with the investigative discourse of the subject (see Box 6.2). You should get the chance to question currently held theories and paradigms, become familiar with terminology and discuss new ideas with those who work around you, including research staff and postgraduate students. The project environment, whether a lab, a library, a field location or a production line, will give you an insight into how research is conducted and how progress is made. It should also allow you to take responsibility for your own progress and make risk-free mistakes.

With a little thought, these criteria can be applied not only to overtly experimental or analytical projects but also to those concerned with the interpretation, presentation or application of

Box 6.2   Real research?

Consider two project scenarios.

A	B
Your project involves analysis of a large number of animal blood samples, obtained by your supervisor during evaluation of a drug.	You work closely with your supervisor on a bioinformatics study which potentially elucidates a tiny aspect of design of an anti-inflammatory drug.
Your task is to assay hormone concentrations in the samples, accurately and reliably, using a routine lab protocol. You must then summarise the results with descriptive statistics and present them to your supervisor.	The study is built around a clear, testable hypothesis. This was formulated by you, with guidance from your supervisor, following a detailed appraisal of research literature.
The reason for the investigation is fully explained and you are given key background papers to read. However, you have no control over the design of the study, no flexibility in how the samples are analysed and little opportunity to interpret the data.	The investigation uses data from a database, made available by a research institute. You create some simple search tools, extract a subset of data and then analyse and interpret the results. You reach a conclusion about the importance of a particular element of molecular structure in drug–receptor interaction.
From your diligent and careful work, a set of data is generated which your supervisor says is of value to his research. It will be included in his next research paper with an acknowledgement of your contribution.	Unfortunately, towards the end of the study you discover that the research institute has now retracted part of the dataset because it was from an unreliable source.

- In scenario A, you did careful, productive scientific work in a self-reliant way but its value to you was minimal, other than learning a routine lab technique. You were acting essentially as a technician and might even feel that some exploitation had taken place.

- In scenario B, you acquired essential investigative skills, obtained a deep understanding of bioinformatics research and molecular structure and took intellectual responsibility for a scientific enquiry. Unfortunately, the results turn out to be of no value in drug design.

The distinction here, of course, is in your autonomy and in the educational value of the research experience. The project in B, though it turned out to be totally unproductive for reasons beyond your control, was real research and highly valuable to you.

In contrast to A, the usefulness of the experience in B was independent of the outcome. It would have been exciting to have made a discovery, but ultimately that was much less important than engaging with the process and delving deeply into an important aspect of pharmacological research.

knowledge. In these situations, novelty comes from observing and reflecting on actions and their effectiveness, or from measuring the success of a new idea or product.

6.5.2 Does it have a clear structure?

The most productive student projects are well organised and have a well-structured programme of investigation: the purpose is clear from the outset and there are well-defined goals. These are characteristics of any good research but they assume a particular urgency for student projects: time is very short and there will be little scope for changes of direction. Your project needs to be well thought out in advance, and you need to see a clear target ahead of you. If you feel that objectives are vague, ask your supervisor for further advice and guidance. Setting objectives, together with a timeline, should be an integral part of project planning; good research cannot take place until they are firmly in place. You may be asked to prepare a mini 'grant application' to ensure that all aspects of the work have been carefully considered. There are some guidance notes for tutors in Section 6.10.2, Project planning and mock grant proposals.

In all types of project there is much to be gained, including objectivity, by envisaging all possible outcomes. This is actually part of the Scientific Method, as delineated in Box 6.1, because the process of hypothesis testing requires all possible results to be anticipated and accounted for. In experimental projects, especially where a statistical analysis will be required, it is usually straightforward to imagine every type of result, including those which are very unlikely. In fact, this is essential in setting experimental conditions (controls, replication etc.). The purpose is not to influence the outcome. It is to pre-empt problems as much as possible and ensure that there will be no nasty surprises when the results eventually come through.

The same principle applies to other projects: you should know what possible qualities and characteristics the result or product might have, even though you may not yet know exactly how it will look. For example, when designing a new product or developing an educational brochure, besides envisaging what it will look like you need some way of knowing if it will be effective. Facing up to the possibility that it may not work as you hoped will help in thinking through the characteristics it needs to possess. With a literature review (see Case study and Prologue), where the aim is to create something (your conclusions) greater than the sum of the parts (the research papers you have studied), it is possible to begin the process by postulating the sort of links and relationships you might find and then go looking for them.

6.5.3 Case study   The value of a literature or data review

Consider the project on elephant reproduction described in the Prologue for this chapter.

This investigation was entirely literature based, yet the student was able to use existing knowledge and published information to create a new, and ultimately testable, hypothesis.

Having immersed herself in various types of complex information, she was able to make new linkages between existing ideas and take a significant intellectual leap. In the final report, the whole was definitely more than the sum of the parts.

Neither the student nor the supervisor could have predicted this outcome when the project began: all they knew at the outset was that there was a problem to be studied. In fact, the new idea did not emerge until the student sat down to write her report and began to assemble what she had found out.

This example also illustrates how ideas may evolve over the course of a project. Projects need to be planned and hypotheses set, but student and supervisor have to be prepared for surprises and to adapt the work as more information emerges.

6.5.3.1   Systematic review

Systematic review is a standardised approach to analysing the outcome of multiple investigations. Such reviews include the meta-analyses which are common in the medical, healthcare and pharmacological sciences.

The principles of systematic review can be applied in any situation where general conclusions are sought from a range of primary studies. The aim is to reach meaningful and reliable conclusions and to avoid errors in interpretation.

Consider using the approach of systematic review if your project involves the appraisal and synthesis of similar types of data from a large number of published studies.

Research is an untidy business. Scientific papers falsely portray studies as a series of step-by-step investigations, each perfectly planned, well controlled, highly organised and logically arranged. In reality, science never works like that. It can be frustrating and confusing to find that you have constantly to jump between reading literature, designing experiments, writing reports, learning methods and analysing data. The work can appear muddled and nothing ever seems to get finished.

The solution to this problem is a structured approach. Even if elements of the work need to be done out of sequence, you should have a clear framework into which they can be securely fitted. This will make it feel as if progress is being made. A jigsaw puzzle is built, not by starting at one edge and moving smoothly towards another, but by looking for pieces and filling in gaps until the overall picture emerges.

When you discuss this with your supervisor, he/she should be able to reassure and guide you, recognising and taking account of your lack of experience. With their help, you will be able to design a strong but flexible structure which can accommodate unexpected issues and tactical adjustments. In fact, one of the most helpful things your supervisor can do is to show you, from time to time, how everything you have done fits into the overall plan. You can discuss how much time to spend on different pieces of work and how to balance your efforts. Use the plan to review progress at regular intervals as the work proceeds. This will reassure you that progress has been made, even when everything seems to be in a muddle.

6.5.4   Are risks identified and responsibilities defined?

Research is, by definition, uncertain. It involves taking risks and being prepared for surprises. You might find this motivating but also worrying. Your project may have been sold to you on the basis of its excitement and the chance of discovery, yet you will be acutely aware of the need to perform well and score high marks. Any project should include some element of fascination with the unknown, but you will quickly become anxious if things are too risky and you sense a loss of control over the results of your work. You also need confidence that the project has been established for the right reasons. The idea for a particular investigation may well have arisen in a research context, yet it needs to be part of your degree programme and has to have education rather than research as its principal objective.

Your supervisor should be able to reassure you that the work will be valued as an essential element of your studies and also that the practical and intellectual risks have been anticipated and properly controlled. It is their responsibility to make sure that unpredictable elements of the investigation do not outweigh the chance that you will be able to see the work through to a satisfactory conclusion. You should both be clear about the project's priorities. At the end of the process, you will be assessed on your performance and it would be unfair for this to depend on elements beyond your control.

The planning stage of a project should consider its requirements in terms of finance, time, data access, permissions, safety and risk assessments, training, the help of others and other resources. All these things must be overseen and accounted for by the supervisor, even if you and others are directly involved in dealing with some of them. Once the project is underway, its progress must not depend on decisions or questions decided elsewhere. Neither should you feel that you are being asked to take on any undue personal responsibilities.

Although projects always have an identified academic supervisor, day-to-day management is often delegated to other staff members (for example, post-doctoral researchers or experienced lab staff). It is not unusual for a non-academic member of a research team (a post-doc, or even a post-grad) to be asked to guide or supervise work or provide a specific element of training. Your project may also rely on the collaboration and cooperation of others outside the department. Students who work externally (on placement with a commercial organisation, in another lab or with an external specialist) will have an identified individual outside the university who agrees to look after them and facilitate the research. However, in all these cases, the academic (university) supervisor retains overall responsibility and must ensure that suitable progress is made. This chain of responsibilities should be made clear to you when you start your project.

Be especially careful if you find yourself working under the guidance or mentorship of a research student. However skilful and willing to help, they have their own research to do, have limited time and must work to their own educational imperatives. Even the most helpful and enthusiastic post-grad can quickly become overwhelmed if your work throws up unexpected problems. They may also lack the experience and judgement of a more established scientist or academic. Make sure your supervisor keeps close account of what's happening.

A key item for discussion with your supervisor is the amount of time you have available. Some departments allocate project work to a specific number of weeks when other studies have ceased. More commonly, the project runs for a longer period but in parallel with other studies. Working outside term time may be appropriate or even necessary, particularly for fieldwork or seasonally dependent experimentation, but this needs to be carefully managed and you should understand and agree to such obligations from the outset.

As the project starts, you will be expected to read literature, summarise it in writing and get to grips with background theory. Depending on the topic, there may also be pressure to start making field observations, generate data or perform experiments. There is often insufficient time for everything to be done at once. In particular, it is difficult to read background material to the required depth or breadth before the investigative work begins. You may have to take on trust the legitimacy of some elements of the investigation and the reliability of the methods you are expected to use, at least until you have had a chance to get to grips with them yourself.

It is inevitable that you will have to organise your time with great care, to fit everything in and ensure that other studies are not neglected. In fact, you will have to multitask at several different levels, and time management might be the most important skill you acquire.

6.5.5   Will I be successful?

The success of your project may be measured in at least three different ways: the mark you receive and the contribution this makes to your overall degree; the skills you acquire and the way you exploit these in the future; and the way the whole project experience contributes to your personal development as an individual and as a scientist.

At the outset of your project, you should be given details of the criteria against which your work will be assessed, along with the various weightings (marks) applied to the research activity itself, the written report and other presentations which you may be required to make. These matters are decided by individual universities and departments, but need to be made explicit and must be fairly applied.

Your supervisor and anyone else involved in the assessment will mark your written dissertation but may also have to make judgements about your performance during the research project. They may have to make some numerical evaluation of abstract personal qualities such as your 'independence', 'enthusiasm', 'intellectual ability', 'organisation', 'perseverance' and others. Marking the written work is relatively straightforward. Assessing your ability as a researcher is much more difficult. Your supervisor will have his/her own view but may seek the opinion of others you have worked with. They will each have seen different aspects of your work and will have their own expectations of how a student should perform. As a student, you need to know that you will be assessed as objectively and reliably as possible.

As we saw from the two scenarios in Box 6.2, and the range of project types available (Table 6.2), differences between topics and the uncertainties of research mean that personal performance does not necessarily have anything to do with the 'success' of the work. By the same token, the dissertation needs to be marked on its completeness as a documentation of the investigation, whether or not the result turned out to be exciting, dull or inconclusive.

So, the question for your assessors is: how can such assessments be made fair? The answer is to focus on *process*, not *content*. In other words, you should be marked on what you did and how you did it, not on the outcome in terms of new knowledge or discovery. You should see that your efforts are properly rewarded, but by the award of marks which reflect your intellectual qualities and progress through the research rather than the result of the investigation.

Of course, your fellow students are undertaking projects too. They are all doing different things and you probably compare notes on progress with some of them in the bar or over a coffee. It is hard to judge whether everyone is having a similar experience or whether some projects are easier or more difficult than others. This is particularly so if some are doing lab work, while others are working with library materials, using datasets or are out on placement.

In fact, no direct comparison is possible. It is especially hard to compare rates of 'progress'. Some projects need a lot of preparation and design while others require an immediate focus on data gathering. For some, methods are already up and running, whilst for others the development of a new technique may be the prime aim. Some will be on topics for which there is a vast body of existing literature while others may be opening up new ideas with little previous material to work from.

You will therefore want to know that all projects have equivalent educational worth. Looking through the project outlines and summaries at the time of project selection, you may have felt that some projects were likely to be more difficult than others. By and large this should not deter you; instead your choice should be based far more on your own interests and preferences than any perceived difference in the apparent degree of difficulty between projects.

For reassurance, bear in mind that it is not in the interests of supervisors to offer unrealistically difficult research projects. They have no pre-existing knowledge of how well individual students are going to perform as researchers. It makes little sense for them to set up projects which can only be realistically tackled by exceptional students. It would also be wrong to allocate the toughest or most exciting research work to students who have performed best in earlier parts of the course, whilst leaving simpler material for those of apparently lower ability.

Why is this? There are two main reasons. First, a student's past academic performance is not a reliable guide to their ability as a researcher: all must have the opportunity to shine, to find new abilities and to surprise their teachers. Research is not about knowledge or exam technique but about the ability to carry out a reliable, scientific investigation. The second reason is that 'favouritism' must not be allowed to influence the provision of support or the allocation of resources.

This is an aspect of the bargain you strike with your supervisor when you agree to work together (Luck 1999, 2009). They can expect you to make the most of the opportunity on offer, but in return you can expect them to provide all the facilities you need, and give you unqualified support. Having said that, if your own performance (skill, insight, critical abilities, effort etc.) makes a difference to the outcome, this must be reflected in the mark you get. Most supervisors and markers will approach these matters with fairness and integrity and will not be averse to discussing their expectations with you. It is well worth finding out what you need to do to get a good mark.

The example in the Prologue (see also the Case study) illustrates a characteristic but rather mysterious and exciting feature of the research process. When given the chance, the human brain, in both its conscious and subconscious parts, has a strange ability to make unexpected associations between previously unrelated concepts. This seems to happen over time and probably requires a significant period of immersion in ideas and information. This is one advantage of being able to spend time on an investigation: it allows you to absorb information at a deep level and gives your creativity a chance to mature and flourish, like wine in a cellar (see also Section 1.6, Incubation, in Chapter 1).

As a researcher, you can try to immerse yourself in your project and give this mysterious process a chance to happen. You can rest assured that if you do this, your project will take on a life of its own. Your supervisor will be delighted and impressed and your assessors will give you the recognition you deserve. This has nothing at all to do with the difficulty of your project or the precise nature of the topic. It just needs the right approach and an enquiring mind.

6.6 Working in groups

Some departments offer projects to groups of students. Science is a social activity and research nearly always involves collaboration. This takes place between close colleagues and also, on a much wider scale, between labs and across departmental, institutional and national boundaries. Thus, working as part of a team is realistic and representative of how most research takes place.

Establishing group projects has advantages for supervisors and departments, especially where there are large classes to deal with. It allows investigation of larger research questions or those needing more pairs of hands, or topics with multiple components. Training in essential techniques can be provided more efficiently to a group, and supervision can be turned into an interactive tutorial-type activity.

The big advantage to students is that they can bounce ideas off one another in a way that is generally not feasible with disparate, individual projects. The success of a collaborative project depends particularly on the group dynamic. If your group gets on well together, you may benefit from each other's insights and get a greater sense of progress than is possible by working alone. As with any other social interaction, success depends on the personalities involved, effective leadership within the team, and the levels of cooperation and mutual support which are achieved.

There are some risks to be aware of if you find yourself in a research team of this kind.

(1) Each member of the group should be able to identify their own part of the work or their specific contribution to the project, so that they can produce an individual write-up and be open to individual assessment. As a student, you must eventually feel that you can take full responsibility for some well-defined part of the study, irrespective of your wider engagement with the group.

(2) Each student needs to experience all aspects of the research process (reading literature, setting hypotheses, designing investigations, carrying out the work, gathering and analysing data, interpretation etc.). Dividing the work equitably between team members, but making sure that the total experience is shared, can be difficult.

(3) Weaker students inevitably rely on stronger ones and it can be hard to sort out who has made the effort. Conversely, stronger students may feel constrained or disadvantaged by working with weaker or less committed ones. They might legitimately feel that their chances have been compromised, just by an accident of association.

(4) Individual students naturally work at different speeds and have different aptitudes. This may not matter too much in a well-defined, highly structured practical class but it can have devastating effects on progress in a complex research project.

The balance of pros and cons for group research reflects the unique circumstances of the final-year project. It is intended to be your individual opportunity to experience and take part in real research. It often carries a large weighting in the final degree assessment. If you are part of a group, you should feel that the work is being carefully managed and that your own chances of success are not at risk or diminished in any way by taking part.

It might be wise to discuss these issues with your supervisor at the start of the project, to make sure that arrangements for working both together with other students and individually have been properly thought through.

6.7 Writing up

Eventually, you will need to write a report on your project and present it for assessment. Project reports take various forms: many departments ask for a traditional thesis; while others expect something more condensed, in the style of a scientific paper. You may also be required to give an oral or poster presentation (see Chapter 11 for advice about communication skills).

Your project thesis or dissertation will probably be the biggest thing you have ever written, and it's a tough assignment. On the one hand, it needs to be a clear and concise report on the work you have done. On the other, it should give you the chance to reflect, deeply and intensively, on what you have achieved. You will also face the challenge of expressing things in an appropriate style, using terminology appropriate to the subject matter and arranging everything in a large but properly organised document.

Students approach their reports with different levels of confidence depending on their experience of writing, skill with words and perception of how successful their work has been. There are several guidebooks available on academic writing at undergraduate level (some of these are listed in Section 6.13, Additional resources). They indicate how material should be presented and the 'mechanics' of writing, and also consider the more general challenge of facing up to the task of converting ideas into words. However, there are very few books of this kind specifically aimed at bioscience or even

Table 6.4 Guidance on writing and presenting your project report

Style and structure	• Follow any guidelines you have been given by your department or supervisor. • Look at previous examples of student project reports. • Make use of the Scientific Method.
Past and present tense	• Use the past, passive, impersonal voice for literature, methods, results and discussion; for example: *'It has been shown that...'; 'The survey was carried out on...'; 'Concentrations were found to be significantly higher in...'* • Use the present, impersonal voice for general conclusions and statements; for example: *'These results show that...'; 'It is hypothesised that...'*
Citations (names/dates in the text) and References (in a bibliographic list at the end)	• Keep track of everything you have read. • Support every statement you make either with a cited reference or with evidence that you have collected. • You cannot have too many citations in a paragraph. • Record reference details once, and once only, using the required style for citing and listing sources.
Write as you go	• Write up as you go. • Keep an electronic notebook of ideas and results. • Incorporate material from plans and reports.
Get help and take advice	• Use technical tools (spelling and grammar checker), but use them judiciously • Ask for advice on phraseology and style. • Use the conventions (terminology, abbreviations) of your subject area. • Ask someone not involved with the project (a friend or relative) to read your work and make sure it makes sense.
Consider the reader (and the marker)	• Think about how the reader will see your work; don't just think about it from your perspective. • Use short sentences. • Keep explanations simple. • Check for errors as you go along and again before submission.
Length	• Keep to specified length requirements. • Make the report as short as possible but as long as necessary.

science students. For this reason, it is important to get advice and guidance from your supervisor. It is worth finding out, as you start to think about writing, how your supervisor will support you. For example, are they prepared to look at bits of work as you go along, or would they like to see complete sections or chapters? There may well be a departmental policy about how much help they can provide, and you should find out if this is the case. Use your supervisor's expertise and experience as effectively and productively as you can.

Table 6.4 offers general guidance about writing and presentation of material in a project report. The information is generally applicable in the biosciences and should prove useful if you have no other source of advice. However, in the first instance you should always follow any guidance provided by your supervisor or department.

6.8   The possibility of publication

The natural outcome of any research should be the bringing of findings into the public domain. If you find out something new, there should be some way of telling the world about it. There is little point in trying to expand the sum of human knowledge if what you discover remains hidden.

Professional scientists achieve this by writing research papers, presenting their work at conferences or by offering their results for commercial exploitation. Most academic staff establish their reputations by developing a list of papers, patents or books, and publication is usually a contractual obligation associated with research grant funding.

Unfortunately, there are scarce opportunities for undergraduate students to publish their results, and to that extent your research experience may turn out to be incomplete. Your final report is most likely to be a personal document which you submit for assessment, rather than something intended for publication. The copy held by your department may in theory be available for public access, but is most likely to remain on a shelf or in a drawer.

On the other hand, if you could generate some sort of visible outcome from your project this would really help in terms of job applications and career prospects. It would maximise the impact of your effort and give it the recognition it deserves (Healey and Jenkins, 2009).

Not everything that students do is of a quality suitable for public exposure, but it is surprising how much of it is. Some gets incorporated into larger projects underway in the supervisor's research group and may end up being published by this route. Occasionally, where research has been commercially sponsored, the results may belong to a company. Such research is intrinsically valuable, even if public access has to be restricted.

A number of universities have developed innovative ways of bringing student research to public view. It would be worth having a look at some of these (for example: *Biolog-e, BURN, Origin* – see Section 6.13, Additional resources) and finding out if your department offers this possibility.

There is also a national journal of undergraduate research in biosciences called *Bioscience Horizons* (see Additional resources). This publishes fully citeable research papers, individually authored by students, which have been submitted for independent expert review. The quality and extent of some of these papers is truly remarkable. They illustrate what undergraduate students can achieve and show that much of their research is first-rate science.

6.9   How you can achieve your potential during final-year project studies

- Be aware of the range of formats for projects offered by your institution, and ensure you choose the right one for you.

- Apply the criteria, outlined in this chapter, that will help you decide whether or not a specific project is likely to provide a valuable learning experience.

- Fully understand the Scientific Method, and ensure it is applied throughout your project studies.

- Carefully note the approaches to be taken when presenting your work orally, in poster format or as a dissertation.

- Whenever possible use your initiative and be creative, but remember, your project supervisor is there to help you: when in doubt you should never be afraid to ask for his or her advice.

- Discuss with your supervisor the possibility of publication of your project results.

6.10   Tutor notes

6.10.1   Characteristics of good research projects – supervisor's checklist

Table 6.5 extends the ideas in Table 6.3 to assist supervisors in establishing desirable criteria for undergraduate project ideas. The emphasis is clearly on educational validity, rather than the research components of the work, as a reminder that projects are principally part of the degree programme. They are not just a way of increasing research output, even though they may sometimes do this too.

Several of the criteria in Table 6.5 are offered as a direct response to the pressures of increasing class sizes and diminishing or limited resources. The material in Tables 6.3 and 6.5 might be used in giving guidance to students approaching decisions about which project to select, getting them to think positively about what they might hope to achieve. The criteria outlined could also be adapted for use in project guideline documents, perhaps combined with material from Tables 6.1 and 6.4, Box 6.1 and other publications.

Table 6.5   Supervisor's checklist

Does this project provide the right educational experience?
Is it an appropriate part of the degree programme?

Does the project have intrinsic educational validity?	• Is the project being offered principally for educational reasons? • Does it fit with the degree programme? • Is it appropriate for the level of study? • Are there opportunities for development of skills? • Is there a structured programme of supervision?
Are the intellectual risks acceptable?	• Is the proposed work soundly based in existing knowledge? • Are all relevant literature and data available? • Will results which I consider undesirable nevertheless be of value to the student?
Have all resources been anticipated and secured?	• Is funding in place and are all resources that might be required available? • Is the support that may be required from other staff (e.g. post-docs or postgraduate students) definitely available? • If the project involves teamwork, will individual performance be independent of that of others?
Can the project be objectively assessed?	• Is the project assessable on process, not content? • Is the project clearly assessable using the department's assessment criteria? • Does the project allow opportunities for exceptional performance, and can this be recognised and rewarded?
Does it offer cohort comparability?	• Is the intellectual challenge and technical difficulty comparable with other projects? • Group work: can students take individual ownership and be assessed in individual achievement?

Of course, simply meeting these criteria is not by itself a recipe for research success, and cannot be a substitute for good supervision. A number of the key points of guidance for students, discussed in the main text, can be reiterated from the supervisor's perspective:

- A research project in biosciences is part of a student's education as a scientist. It should therefore be a properly constructed scientific investigation, offering familiarity with the Scientific Method and the opportunity to find out something new. It should also provide clear opportunities to achieve understanding, develop skills and demonstrate aptitude. Indeed, the word 'education' has its root in the Latin word for 'leading' or 'drawing out'; there can be no better opportunity for true education than through guided personal research.

- The actual link between what goes on in a project and its value for education and training is often hard to define. Discoveries can appear modest or even trivial in the grand scheme of things, but still be the result of hard work, imagination and intense personal commitment. Any kind of project should provide the opportunity to think creatively and examine a problem in depth, and to experience feelings of excitement and wonder.

- The criteria suggested in Tables 6.3 and 6.5 are in many ways idealistic, and some are controversial. For example, many academic staff feel that 'real research' is not always a practical possibility, especially where there are large classes or very limited resources. Some also believe that a university science course, even if not vocational, should prepare students for distinct areas of employment. They therefore prefer to put the emphasis of project work on the development of practical or other transferable skills (James, 2009; Bevitt *et al.*, 2010). Nevertheless, in the view of this author, it should still be possible to create a legitimately investigative environment such that the student experiences a real enquiry into the unknown.

- The research project usually carries a high weighting in the final assessment. All students on the course must have an equal opportunity to demonstrate their abilities. It would be quite unfair if some projects were more 'valuable' than others or only capable of successful completion by exceptional or outstanding students. Fundamentally, all students must have an equivalent educational opportunity.

- Marking is always a subjective activity, but well-defined learning objectives and clearly expressed guidelines are the keys to fairness in both expectation and assessment. Assess the process, not the content, and make sure that students understand how marks are awarded.

- Group projects are efficient and allow students to develop valuable team-working skills. However, if such skills are to be defined as a learning outcome, then all students in the year group must have an opportunity to develop them, not just those doing group research (Hejmadi, 2008; Saffell, 2009).

- Even if teamwork skills are a defined outcome of group research, it must still be possible to assess students individually: their individual marks must not depend on the performance of others.

6.10.2 *Project planning and mock grant proposals*

Experience suggests that many students find project planning rather difficult. It requires a mental discipline and foresight which they may not yet have developed, especially if their experience of science so far has been largely passive and observational. Some departments get students to plan properly and think ahead by asking them to prepare a mini grant application. This can be based, for example, on a research grant application form of the type used by the Biotechnology and Biological Sciences Research Council or the Medical Research Council. Typically, the application will ask for a justification of the research based on published literature, together with clear hypotheses, outline methods and procedures for analysis. Resources and timescales can also be anticipated and planned for. This forces the student to think the project through and start to ask some of the questions they will be faced with when the work begins. For the supervisor too, it can be an excellent way of deciding whether a project has been properly thought out.

6.11 Acknowledgements

Thanks to Anne Tierney for case study material and David Adams for valuable discussion.

6.12 References

Bevitt, D., Baldwin, C., Watts, C., Ferrie, L. and Calvert, J. (2010) Can you tell what it is yet? An investigative approach to final year project work. Presentation given at *Final Year Projects: Maximising the Learning*, HEA Centre for Biosciences, University of Newcastle, May 2010. Available at: www.bioscience.heacademy.ac.uk/ftp/events/newcas130510/bevitt.pdf. Accessed April 2011.

Healey, M. and Jenkins, A. (2009) *Developing Undergraduate Research and Enquiry*. York: The Higher Education Academy.

Hejmadi, M. (2008) Mentoring scientific minds through group research projects: maximising available resources while minimising workloads. Leeds: HEA Centre for Bioscience. Case study, available at: www.bioscience.heacademy.ac.uk/resources/guides/studentrescs.aspx. Accessed April 2011.

James, H. (2009) Analyses in biology: an analytical alternative to traditional research projects. Leeds: HEA Centre for Bioscience. Case study, available at: www.bioscience.heacademy.ac.uk/resources/guides/studentrescs.aspx. Accessed April 2011.

Jenkins, A., Breen, R., Lindsay, R. and Brew, A. (2003) *Re-shaping Higher Education: Linking Teaching and Research*. London: Routledge.

Luck, M. (1999) *Your Student Research Project*. Aldershot: Gower.

Luck, M. (2009) *Student Research Projects: Guidance on Practice in the Biosciences*. Leeds: HEA Centre for Bioscience.

Luck, M. and Wagstaff, C. (2003) The Scientific Method (www.bioscience.heacademy.ac.uk/ftp/Teaching-Guides/studentresearch/Scientific_Method_Luck_Wagstaff.pdf). Accessed April 2011.

Ryder, J. (2004) What can students learn from final year projects? *Bioscience Education* 4,2. www.bioscience.heacademy.ac.uk/journal/vol4/beej-4-2.aspx.

Saffell, J.L. (2009) Group research projects: a framework for providing research experience for students. Leeds: HEA Centre for Bioscience. Case study, available at: www.bioscience.heacademy.ac.uk/resources/guides/studentrescs.aspx. Accessed April 2011.

Taraban, R. and Blanton, R.L. (Eds.) (2008) *Creating Effective Undergraduate Research Programmes in Science: The Transformation from Student to Scientist*. New York: Teachers College Press.

6.13 Additional resources

There are many other sources of guidance on how to approach your project and maximise the opportunities it presents, particularly in the context of your other studies. You might find the following helpful:

Allison, B. and Race, P. (2004) *The Student's Guide to Preparing Dissertations and Theses* (2nd Edition). Routledge Study Guides. London: Routledge.

CETL-AURS. Centre for Excellence in Teaching and Learning in Applied Undergraduate Research Skills website. University of Reading. www.engageinresearch.ac.uk/. Accessed April 2011.

Jones, A., Reed, R., Wyers, J. and Martini, F.H. (2007) *Practical Skills in Biology* (4th Edition). Harlow: Pearson Education.

McMillan, K. and Wyers, J. (2007) *How to Write Dissertations and Project Reports*. Harlow: Pearson Education.

Race, P. (2000) *How to Win as a Final Year Student: Essays, Exams and Employment*. Buckingham: Open University Press.

For guidance on **experimental design** and the **Scientific Method**, see:

Barnard, C.J., Gilbert, F. and McGregor, P. (2007) *Asking Questions in Biology: A Guide to Hypothesis Testing, Experimental Design and Presentation in Practical Work and Research Projects* (3rd Edition). Harlow: Pearson Education.

Luck, M. and Wagstaff, C. (2003) The Scientific Method (www.bioscience.heacademy.ac.uk/ftp/TeachingGuides/studentresearch/Scientific_Method_Luck_Wagstaff.pdf). Accessed April 2011.

Information on **systematic review** as an approach to literature analysis can be found at the Cochrane Collaboration website: www.cochrane.org/information-researchers-and-authors#more. Accessed April 2011.

The following sites **publish** or provide a **showcase** for research by UK undergraduate students (accessed April 2011):

Biolog-e. University of Leeds. www.biolog-e.leeds.ac.uk/Biolog-e/index.php.

Bioscience Horizons – publishes single-author research papers and reviews by UK undergraduate students. All articles are subject to stringent quality assessment by experts in the field, as they would be for any other scientific journal. Articles cover the full range of subject areas represented by UK bioscience departments, from molecular medicine to environmental studies. They are fully citable and freely accessible online from the moment of publication. http://biohorizons.oxfordjournals.org/.

BURN (Biosciences Undergraduate Research at Nottingham). University of Nottingham. www.nottingham.ac.uk/burn.

7 Maths and stats for biologists

Dawn Hawkins

7.1 Introduction

7.1.1 Aims

This chapter is all about empowering you to acquire the mathematical and statistical skills you will need as a twenty-first century bioscientist. Once you have interacted with the material in the chapter you should:

- Understand the importance of maths and stats in the biosciences.

- Be able to implement strategies to broaden your skills base in maths and stats.

- Feel more confident about using maths and stats to help you fulfil your potential as a bioscientist.

7.1.2 The relationship between maths and stats

Although this chapter is entitled *Maths and stats for biologists*, strictly speaking statistics is a branch of applied, or applicable, mathematics. Computer science can be viewed as an example of another branch of applied mathematics. The main branches of 'pure mathematics' include geometry (the maths of space) and analysis, including calculus (the maths of rates of change). Geometry and calculus are useful, for example, in studying animal ranging patterns and the speed of biochemical reactions, respectively. All of these elements are important for biologists but the emphasis given to statistics in the title of this chapter reflects the universal importance of this branch of maths in biology because nowadays, for reasons I shall explain, virtually no research in the biosciences can be published without some form of statistical analysis.

In the rest of this chapter I use the terms 'maths and stats' together a lot, but in the instances where I refer only to 'maths' this is usually for the sake of succinctness and should not be taken to imply the exclusion of 'stats'.

Effective Learning in the Life Sciences: How Students Can Achieve Their Full Potential, First Edition.
Edited by David J. Adams.
© 2011 John Wiley & Sons, Ltd. Published 2011 by John Wiley & Sons, Ltd.

7.2   Motivation – this chapter is important!

I thought of asking the editor if I could rename this chapter 'Read me – I'm important', as getting people to even look at a chapter entitled 'Maths and stats' is likely to be a bit of a hurdle. That I didn't, and that you are still reading, I hope indicates that you have some level of motivation, or a vague curiosity at least, which is a start. However, my guess is that you don't yet appreciate how crucially important maths has been, is, and will increasingly be, for biology, and therefore what an important tool it represents for you as a bioscientist. I want to instil in you an appreciation of the importance of maths because without it you won't have the fundamental requirement for effective learning: the *motivation* to learn.

7.2.1   What no functional biologist should be without

So, in a nutshell, why, as a biologist, do you need to deal with maths and stats? The short answer is simple: you will not be able to **be** a bioscientist if you do not, and your employment prospects will be severely compromised (Tariq *et al.*, 2010). As I explain below, acquiring skills in maths and stats is essential in facilitating three key processes: participation, communication and evaluation.

7.2.1.1   Participation

Without some basic skill and understanding in maths and stats you can be reasonably knowledgeable about biology, but you won't be able to participate in the generation of new knowledge. Think about the biological information that you enjoy learning about: where did it come from? It is one thing to passively acquire factual knowledge, and quite another to contribute to the generation of that knowledge through the development and testing of biological theory. To do the latter you will, at the very least, need to have enough awareness and skill to work with others to generate and analyse data and models. Statistical techniques, for example, are a vital tool in analysing data and answering biological questions, and you really will need to be able to conduct basic statistical analyses yourself. It is true that research biologists often work in teams that include statisticians, but even then each member of the team needs a basic level of understanding to be able to work effectively with the others. In this case ignorance will certainly not be bliss: it will lead to isolation, frustration and unemployment.

7.2.1.2   Communication

Once you have participated and produced some findings, you will need to communicate these to other scientists; for example, the person marking your undergraduate assignment or the person reviewing your scientific paper, or it could be other members of your research team. In any event you will need to know the basic language and formats used to effectively convey knowledge and information to others. Most frequently this applies to the Methods and Results sections of publications, including undergraduate reports and dissertations. For a Methods section you might, for example, be required to report concentrations and dilutions of solutions in ways that others can quickly and easily understand. Statistical analyses are a common feature of Results sections and you need to know, for example, the amount of information, contained in a computer printout, that should be incorporated in your report: typically you should include the name of the test, the degrees of freedom and/or sample size, and a P value.

7.2.1.3   Evaluation

If you communicate your results clearly then others should find it easy to evaluate your work. You also need a reasonable grounding in maths and stats if you are to evaluate the work of others. Scientists are fallible, certainly much more than the popular image of a scientist suggests. The ability to critically evaluate all aspects of research is a key skill for any scientist. The more maths and stats skills you have, the less likely you are to be duped!

7.2.2   Maths that has changed, is changing and will change biology (and vice versa)

For over 1 400 years it was believed that the human circulatory system involved blood flowing backwards and forwards in arteries and veins! Putting together some careful anatomical observation with a bit of nifty but basic maths, William Harvey finally put paid to this idea in the early seventeenth century (Cohen, 2004). Using basic maths, he estimated how much blood flowed through the system based on the size of the heart chamber (left ventricle) that pumps blood out of the heart and around the body, and the frequency of pumping (the heart rate). He based the former on measurements of the dissected hearts of dead people, and the latter on measurements of live individuals: he determined their pulse rates. His answer was a flow rate of 27 litres per hour. He also measured the total volume of blood in dead bodies and found it was only around 5.5 litres. The two numbers did not match and there was clearly no way 27 litres could be flowing backwards and forwards in two separate systems.

Harvey therefore postulated that the arteries and veins must, in fact, be joined as part of a continuous system with the 5.5 litres circulating rapidly around the body under the control of the heart. He was, of course, absolutely right, and the existence of capillaries was confirmed a few years later. With some measurement, some geometry (to work out volume occupied by blood) and some basic arithmetic, he completely changed our understanding of how our bodies work.

There are many other examples. Jungck (1997) identified 'Ten equations that changed biology', covering: genetics and evolution (integration of genetics and evolutionary theory into neo-Darwinism, mapping of genes and construction of phylogenetic trees); ecology (predicting rates of population change when species compete, and the relationship between the size of an island and the number of species on it); and cell biology and biochemistry (the rate of movement of chemicals across the cell membrane, and the rate of enzyme-controlled reactions). He also pointed out that, without mathematical equations, modern medicine would be without its CAT (computer-assisted tomography) scanners and other non-invasive ways of seeing what the inside of a human body looks like.

Maths has been central to the most profound of all theoretical developments in biology, that of evolution. One nice example is the Hardy–Weinberg principle described in the case study below. Robert May (2004) called the Hardy–Weinberg principle a 'kind of Newton's First Law for evolution'. It demonstrates that inheritance based on particles (genes) can maintain the variation in populations needed for natural selection to work as Darwin proposed.

7.2.2.1   Case study   The Hardy–Weinberg principle

In 1908, Godfrey Hardy, a mathematician from Cambridge University, wrote to the journal *Science* about some work he had heard about by Udny Yule on a medical condition known as brachydactyly (Hardy 1908). Individuals suffering from this condition have relatively short fingers and toes. The inheritance of the condition is determined by a pair of alleles in a simple Mendelian manner, with the allele for brachydactyly dominant over the allele for normal sized digits.

Hardy was alarmed by Yule's claim that if brachydactyly were not being selected against, then the expectation would be for the proportion of brachydactyly alleles in the population to increase until there were three individuals suffering from the condition for each normal individual. Hardy thought it was pretty obvious that this would not happen. He supported his comments with a few general mathematical equations which you can find explained, in modern terms, in most introductory genetics textbooks (for example, see Thomas, 2003).

By way of illustration, Hardy applied his equations in a population where one person is homozygous for brachydactyly for 10 000 individuals homozygous for the normal condition. He showed that in the next generation the proportion with brachydactyly will approximately double but after that there will be no tendency for the proportion to change. To quote Hardy: 'In a word, there is not the slightest foundation for the idea that a dominant character should show a tendency to spread over a whole population, or that a recessive should die out.' He was a bit disparaging that biologists themselves hadn't noticed this: 'I should have expected the very simple point which I wish to make to have been familiar to biologists.' However, at least one biologist was aware of the principle: a similar point was made independently, in the same year, by a German medic, Wilhelm Weinberg. The law is therefore known as the Hardy–Weinberg principle.

The Hardy–Weinberg principle describes the relationship between the frequency of alleles and the frequency of genotypes in the form of a simple equation, and therefore allows one to be predicted from the other. Furthermore, it can be used to demonstrate how allele and genotype frequencies in a population can remain the same from one generation to the next unless, for example, mutations occur. It holds, provided population is large and mating is random and there is no selection. That the Hardy–Weinberg principle demonstrates how variation can be maintained, despite the presence of dominant alleles, is profoundly important in the context of Darwinian theory.

Another area of maths that has had a great impact on our understanding of evolution is *game theory*. This is a branch of applied mathematics that looks at what individuals might gain by making different decisions depending on what others around them decide to do. Game theory has been applied to economics for over half a century, and in 1994 John Nash (whose life was featured in the film *A Beautiful Mind*), along with John Harsanyi and Reinhard Selten, won the Nobel Prize for Economic Sciences, 'for their pioneering analysis of equilibria in the theory of non-cooperative games'. Evolutionary biologists, notably John Maynard Smith, were inspired by economists to apply game theory to interactions between animals to better understand, for example, the evolution of fighting strategies: for instance, when it is better to display or retreat rather than enter into a physical fight.

In the words of Joel Cohen (2004; my formatting):

Mathematics can help biologists grasp problems that are otherwise
too big (the biosphere) or
too small (molecular structure);

too slow (macroevolution) or
too fast (photosynthesis);
too remote in time (early extinctions) or
too remote in space (life at extremes on the earth and in space);
too complex (the human brain) or
too dangerous or unethical (epidemiology of infectious agents).

Cohen contends that maths is to modern biology what the microscope was to seventeenth-century biology, but even more so. This comparison is powerful and appropriate: as the microscope revealed previously unseen organisms and structures at the cellular level, so maths reveals previously unseen patterns in data at all levels. If you think about it you realise that modern students of biology need to be as familiar with using maths as with using a microscope.

We live in astonishingly exciting times. Due to technological and computing advances over the last decade or so, there has been an explosion in the rate at which biologists can generate data. Bioinformatics is the emerging area in the biosciences that harnesses the increasing power of computers to manage and look for patterns in biological data, particularly at the molecular level. For the first time the ability of cell biologists, geneticists and biochemists to generate data is outstripping the ability of available mathematical tools to cope with the analyses. For example, the combination of the polymerase chain reaction (PCR) and increasingly rapid gene sequencing procedures means that vast amounts of data are generated from DNA samples. However, it's impossible to progress from these raw data to answering questions like: 'Did modern humans originate in Africa?' without the use of some serious mathematical tools.

On a macro- rather than micro-scale, global level geospatial data on habitats and animal migrations, generated by satellites, GPS and other such gadgets, can only be effectively visualised and analysed by exploiting the appropriate maths and computing power within GIS (geographic information system) software.

The complexity and variability of biological systems is awesome, and mathematicians can help biologists deal with this. Conversely, biology is increasingly stimulating new realms of maths. In some cases the biologists have not just done the stimulating but the developing too. I don't think it's any exaggeration to say that Ronald Aylmer (better known as R. A.) Fisher almost single-handedly created the branch of mathematics that we know today as statistics. Was he a mathematician? No. He was a biologist – a geneticist to be precise – in need of a tool to figure out the basis for the variation in crop yield noted in a large agricultural dataset generated at the Rothamsted Research Station in Hertfordshire. Returning to the thoughts of Joel Cohen (2004), he also argues that biology has replaced physics as the science providing most stimulation for developments in maths.

7.2.3 Why biologists need statistics (or why a sample of one is useless)

Biologists have to deal with complex and highly variable systems. For example, trying to predict who will develop swine flu, when this will happen, and how badly they will be affected, is a difficult job! We can't possibly get the answer to this by looking at just one person. On the other hand looking at all humans is impossible. What we do is take samples. We need statistics to help us summarise what is in our samples and to help us decide if our samples reveal any biologically meaningful patterns, or not.

As the second case study illustrates, we can observe apparent patterns in our data through chance alone. To further complicate matters, when confronted with raw data (or summaries of the data such as averages), humans are very bad at telling the difference between patterns created by chance and patterns that reflect a real biological relationship. Statistical analysis allows us to make complex biological data manageable and to detect biologically interesting patterns in an objective and clearly defined way.

7.2.3.1   Case study   Why we need statistics

Imagine entering a large supermarket and recording two items of data for each person in the store:

(1) The side of the store they are in: left or right.

(2) Their height in metres.

Now imagine calculating the mean height of people in the left and right sides of the store and comparing the means. Ask yourself:

Will these two means be exactly the same?

My guess is that you would be very surprised if they were the same; you may expect them to be similar, but not exactly the same. You are likely to accept that the difference in means has no real significance, has happened by chance (a phenomenon statisticians call 'sample error'), and you would not wish to come up with a hypothesis or biological explanation for your observation.

OK. So now you are in the fracture clinic in your local hospital where only half the patients have received a new drug designed to stimulate bone repair. As the plaster casts are removed you record two items of information for each patient:

(1) Whether they received the drug.

(2) The time (days) required for the fracture to heal.

The data you collect reveal that the mean time required for healing for those taking the drug was 45 days and for those not taking the drug was 48 days. What do you conclude? That somebody is on to something important with this drug? Wrong! This difference could be a product of chance just like the difference in mean height for people on the left vs. the right side of the supermarket. You need some statistical tools to help you reach firm conclusions and to be able to convincingly communicate your conclusions to other scientists.

7.2.4   Tutor notes   Motivating students

Context is crucial: the factor that motivates students most is seeing techniques used in a context that is relevant to their studies. Bioscience students need examples from the biosciences. Even better, from the specific topic they are currently studying, for example ecology, pharmacology, genetics etc. A textbook or website with a biology-specific example is far better at engaging bioscience students than a more general resource.

The *by biologists for biologists approach*: I am not alone in my experience that students are more motivated when taught by staff they know, from within their own bioscience department, than by service teachers supplied by mathematics departments. This also relates, in part, to context, and links to confidence issues addressed in the Tutor notes in Section 7.3.3. The best scenario is a course designed and taught by biology staff with access to a biology-friendly mathematician who can check for mathematical violations!

Enthusiasm is more important than expertise: if teaching materials are prepared thoroughly and are well structured, teachers with very little maths experience can be very effective in passing on knowledge and skills to students. For example, the involvement of postgraduate students as Graduate Teaching Assistants (GTAs) can greatly improve teacher: student ratios in a cost-effective way. A situation like this, in which teacher and student are seen as learning together rather than the former instructing the latter, can be very successful if acknowledged and supported.

7.3 Confidence – you can do it!

By now you should be convinced that if, instead of being just someone who knows a lot of biological facts, you are to be someone who is able to understand, evaluate and contribute to the full breadth of current biological thinking, you will need some maths and stats skills. But maybe suddenly you are thinking 'Oh heck' (or some such similar phrase), 'I'm done for – need to find a different career!!' Not so – read on.

7.3.1 My Maths Mantra

If you have little or no confidence about using maths and stats in biology, here is a little mantra that should help whenever the need arises:

(1) I am not alone.

(2) I don't have to become a mathematician.

(3) If I get the wrong answer it does not mean I don't understand.

(4) I won't lose all my friends.

(5) Yes I can, in fact I do.

I am not alone

If you are feeling intimidated by the need to acquire sufficient maths and stats skills to be an effective biologist, rest assured you are not the only one. 'Maths anxiety' is a widely recognised phenomenon, and amongst the STEM subjects (that's Science, Technology, Engineering and Mathematics), biologists seem to be particular sufferers. Teaching professionals hold workshops on the 'maths gap' – the gap between the maths skills students have and the maths skills they need (Carter *et al.*, 2010).

I've been teaching introductory statistics to first year undergraduate biologists in the UK for over 15 years and I have met a lot of maths-anxious students. More broadly, the phenomenon has been reported at national and international levels, for example by the United States Department of Education (National Mathematics Advisory Panel, 2008).

You are not alone; lots of bioscientists think they don't have any ability in maths, but remember, there are lots of people and resources out there to help you.

I don't have to become a mathematician

For biologists, maths is a tool, a means to an end, not an end in itself. You don't necessarily need to learn lots of equations and formulae. It is a really good idea to be able to do basic calculations and be familiar with a few basic concepts and often-used techniques. However, it is important that you also make it a priority to understand how you can find out what maths can do for you. This means knowing about the resources, such as books and websites (see Section 7.8, Additional resources for examples), that are available, and being able to make good use of them (see Section 7.3.2).

Of course, another way you can find out how to do something mathematical is simply to ask someone. To do this you not only have to have an idea of who to ask but you also have to think about how you will phrase your question so that it will be understood by someone who may not have had any biological training. Once again, this does not require any detailed mathematical knowledge on your part.

You can remain a biologist; you just need enough maths to engage with source material and the people who have more maths than you.

If I get the wrong answer it does not mean I don't understand

Give me my end-of-month bills to add up manually and I'm unlikely to get the same figure twice. This does not mean I don't understand the concept or procedure required, and it doesn't make me enter a downward spiral of doom: I just accept that I make mistakes. I find my students are a lot less forgiving of themselves. For example, if they follow the correct procedure for a t-test and, for example, obtain a value of 2.85 instead of the correct answer of 1.72, they have a tendency to write themselves off in terms of stats. It's quite frustrating.

You must remember that making mistakes is not synonymous with misunderstanding.

I won't lose all my friends

Early in my lecturing career, I was sitting at my desk on campus one day when there was a furtive knock at the door. I looked around and could see one of the students, student X as I will refer to her, from the first-year biological statistics course that I teach, anxiously looking both ways down the corridor. I said 'Come in.' She did. I said 'Take a seat and tell me how I can help.' She didn't respond. She just remained standing about equidistant between me and the door looking troubled in a way I did not recognise. I was used to people coming in with their stats troubles, needing help with a t-test or something, but this was different. I said, 'What's up?' There was a bit of a pause and she looked over her shoulder. Then she confessed 'I'm really enjoying learning about statistics,' rushed to the chair, sat down and waited to be counselled and cured of this worrying affliction.

Now, many years on, I'm used to this kind of thing. A lot of students don't perceive it to be very cool to be good at maths and stats. For women there is an added dimension in that maths can been seen as a somewhat unfeminine subject at which to be good. Such unnecessary barriers to learning need to be made a thing of the past. If you understand how important maths is going to be in the twenty-first century, not just for biologists but for everyone, you will know that being maths savvy, or at least maths tolerant, will rapidly become one of the ways you can make friends and influence people.

You won't lose any friends if you reveal you have some interest and ability in maths, but you might make some new ones.

Yes I can, in fact I do

It is likely that you constantly use mathematical concepts and procedures in your everyday life. For example, even if you aren't very good at basic arithmetic and make lots of mistakes, I'm sure you use procedures like adding, subtracting, multiplying and dividing all the time. You might have thoughts like this: 'I've got five pounds, my bus fare will cost two pounds, do I have enough for a pint of beer when I get there?' You undoubtedly have a working grasp of more complex concepts like probability: 'What are the chances that the bus is going to be on time'? 'Do you think he's likely to ask me out?'

In a historical context the mathematics you perform without thinking too much about it is astonishing. The concept of *zero* didn't reach Western Europe until the twelfth century, but I don't expect you have much trouble understanding or using the value for nothing. In the seventeenth century, nearly 400 years ago, it was a big deal that René Descartes came up with a system that allowed points to be located using two coordinates and enabled distances between these points to be measured. Today you use the same principles when you plot a graph using a horizontal x-axis and a vertical y-axis, and you probably don't think too much about it.

You already have the ability to do maths.

7.3.2 Learning the lingo

How would you feel if I asked you to construct a linear model of distance travelled by an elephant walking at a constant four miles per hour? Not sure? What if I told you that all this involved was drawing a straight-line graph? Does that sound more like something you could do?

We shouldn't be too hard on the mathematicians. We all do it; we create and use a 'lingo' for areas we are especially familiar with. It might be in opera ('that diva is singing an aria': in other words, that well-known female opera singer is singing a solo) or neurobiology ('a major function of the cerebellum is motor control': in other words, the extra wiggly bit at the bottom of the back of your brain helps you move). Not knowing the vocabulary used by people working in an area may mean that perfectly simple statements, explanations and instructions will look and sound like gobbledygook. It's therefore worth putting in some effort to understand the maths lingo.

A good illustration of this is the increasingly frequent use of the word 'model' in biology. For example, you might read about the Marginal Value theorem which is a model for the time animals spend looking for food in a patch in relation to how far they have travelled to get to that patch. Alternatively you might be shown an epidemiological model of the spread of a disease affecting livestock. This may sound a bit scary at first, but a model is simply a description of what a scientist thinks is going on, and the model he or she generates may then be expressed as an equation or as a graph. So, for example, if you plotted a graph of the weights of 100 people against their heights, and drew a line of best fit through the points, you could say that you had generated a linear model of height against weight.

Another potential point of confusion is when different words are used to describe the same thing. For example, the words *response*, *test*, *outcome* or *dependent* before the word *variable* all have essentially the same meaning. This means that you might read an article in which the term 'test variable' is used in a context where you thought the term 'dependent variable' should be used, and this can lead to an unnecessary loss of confidence.

Websites and textbooks frequently have glossaries of mathematical terms, and you should make use of these, but an even better way of getting to grips with the terms and expressions you need to know is to compile your own personal glossary using official definitions and, most importantly, your own interpretations of the terms.

7.3.3 Tutor notes   Building students' confidence

Ideally, in any module involving maths for biologists, enhanced student confidence should be a specific and explicit learning outcome. You can build student confidence by:

- Accepting high average class marks: answers to maths problems tend to be clearly right or wrong. It is better to set questions at a level that the majority of students can get right if they have learned the material and practised the techniques.

- Giving students plenty of practice in basic calculations. Making a mistake is not the same as misunderstanding, but getting the 'wrong' answer repeatedly will certainly discourage students.

- Encouraging conversation and social interaction within class time: this can boost confidence as well as social acceptance of maths and stats! For example, you might ask a few questions, involving maths, of the whole class, and then allow time for the students to discuss the answers amongst themselves.

- Making sure your expectations match current practice: you must be careful not to undermine student confidence by expressing out-of-date ideas about the maths they learned at school. Maths at secondary school level has changed significantly over the years and is continuing to change (Lee *et al.*, 2010). It may be that the students you believe to be under skilled are, in fact, highly skilled, but with a different skills profile.

7.4   Skills – do it!

I am hoping that you have now reached the point where you are eager to expand your maths skills base and are feeling OK about doing so. How do you proceed? Firstly remember the mantra: 'You don't have to become a mathematician.'

It's handy to have a working knowledge of basic arithmetic and statistical procedures. Clearly, for example, it's good to be able to add up, or to calculate a mean, without having to consult a reference book or ask someone how to do it. However, it is also very important that you should recognise when you need help and know how to go about getting it. In this section I am going to describe strategies to help you do this; namely how you can get an overview and find commonalities, plan ahead and choose the right technique, and find and evaluate resources. I will focus, in particular, on statistics, but the principles apply generally to all aspects of maths.

7.4.1 Getting an overview and finding commonalities

The range of maths and stats techniques available to analyse your data is huge and expanding rapidly. Starting out, this can seem more of a hindrance than a help! It's easy to get overwhelmed by the choice. There are two issues: one is being aware of what is available; the other is choosing the appropriate technique for a particular situation. We will look at choosing the right technique in the next section. Here, I will consider how you might get your head around what is available by developing an overview and finding commonalities within the bigger picture.

You should start by developing your own overview: a framework for your learning that can develop and expand as your learning progresses. The format can be a list, but you may prefer a 'mind map' approach as a diagrammatic representation of your thoughts (see Chapter 1). It's a good idea to put these sorts of overviews together leaving space for your own, personal notes and answers to questions.

Whichever format you use, it should be something you can interact with effectively and that can evolve with the expansion of your understanding and knowledge. For example, as questions occur to you, you should write them down on the overview and then seek to answer them through reading or talking to others; you should constantly annotate your overview with the answers to questions along with your thoughts and observations. It is quite likely that at some point it will be appropriate for you to rebuild the whole thing.

As well as trying to get an overview you should also look for commonalities. This approach will radically reduce the amount of learning you have to do. For example all traditional hypothesis testing techniques, such as t-tests, Wilcoxon paired t-tests or Pearson correlations, have four steps:

Step 1: Construct a null hypothesis (see also Section 6.3, Chapter 6, *Applying the Scientific Method*).

Step 2: Decide on the critical significance level, α (typically 0.05).

Step 3: Carry out statistical calculations with degrees of freedom (where appropriate) and accompanying P value (if you are using statistical software).

Step 4: Decide whether to reject or accept the null hypothesis (Step 1) based on the results from Steps 2 and 3.

This approach is described in full in Hawkins (2009). The point here is that, if you know the four steps, then you can see the commonalities between techniques and a lot less learning should be required. The calculations for Step 3 obviously differ between techniques and you can find out more about these as you need them. For Step 1, the null hypothesis is, in general, that there is no biologically interesting pattern (difference or relationship) in the data; in other words that any apparent pattern is a product of chance alone. For Step 2 you decide on a critical significance level. The choice is yours, but a level of 0.05 is typical. Finally, for Step 4, you will make a decision on whether to reject or accept the null hypothesis. If you are using statistical software and have generated a P value as part of your calculation, then you simply compare this value to the critical significance level in Step 2: if the P value is equal to or less than the critical significance level then you reject the null hypothesis. In other words you reject the possibility that the pattern is due to chance alone and thus accept that your pattern is significant.

If you perform Step 3 manually then you will have to compare the value of the statistic that you calculate to a critical value for the statistic that you look up in a pre-prepared table, using the critical significance level from Step 2 and degrees of freedom or sample sizes. The rule typically is that you reject the null hypothesis if the value of the statistic that you calculate in Step 3 is less than the critical value that you look up in a table.

7.4.2 *Planning ahead and choosing the right technique* ————————————

It's a really, really good idea to consider, from the start, the maths and stats skills you will need. In terms of stats, R. A. Fisher is reported to have said as early as 1938, to the First Indian Statistical Congress, that 'To consult the statistician after an experiment is finished is often merely to ask him to conduct a post-mortem examination. He can perhaps say what the experiment died of.' Certainly, choosing the specific statistical technique you plan to use *before* you collect data is an exceptionally good idea, and this principle applies for all work that will involve maths and stats. For example, if you will need to perform dilutions as part of an experiment you should calculate precisely what you will have to do before you start, to avoid unnecessary distractions once experimental manipulations are underway.

Choosing the appropriate statistical technique to use to analyse data is a crucial part of the process of experimental design (for an excellent introduction to experimental design for biologists see Ruxton and Colegrave, 2011). In short, if you wait until you have collected data before thinking about statistical analysis, you risk wasting the time and the resources you have used. When these resources include living organisms then there is a potential ethical cost as well as a financial one. Conversely, if you do choose an appropriate statistical technique early on in the experimental process, this will not only improve the efficiency and quality of your science but will also lead you to an easier, happier life.

So, how do you go about choosing the right technique for you? A number of textbooks and websites offer resources that should help. Typically these are in the form of flowcharts or tables. Some people find it hard to choose a technique prior to generating experimental data. One way round this is to conduct preliminary or pilot work, and produce some 'pilot data' to work with. In addition, or alternatively, you could just make up some 'dummy data'! No, really, this is perfectly legitimate! You can invent dummy data at the planning stage, as illustrated in Worked example 1, and then discard them once you have some real data. This can be a very good exercise that will help you identify variables and think about how you are going to measure them. Looking at previously published work can help you make the data more realistic, but this isn't essential.

An example of how to choose a statistical hypothesis testing technique, using the approach described at the *NuMBerS* and *NUMBAT* websites (see Additional resources), is provided in Worked example 2; the example uses the 'dummy data' generated during Worked example 1. You may find that, at the planning stage, you cannot decide on the precise technique you should use. However, you should at least be able to identify a number of alternative techniques. For example, as you progress through Worked example 2, you will not be 100% confident that real data will fulfil the additional criteria needed to conduct a parametric test rather than a non-parametric one, but you will at least have explored a number of techniques that you might employ when you analyse the real data generated during your experiments.

7.4.3 Worked example 1 Creating dummy data

Task: make up a dummy data set that could be used to answer the question: does bone density differ between adult females over 50 years of age in the USA compared to China?

Things to think about:

- How many samples do I need?

- How big should each sample be?

- What should the data in each sample look like? If the data are in the form of numbers, what sort of values would be realistic?

In this example you will need two dummy samples, one for each country. With regard to sample size, in terms of statistics, the bigger the sample the more powerful your test will be (that is, the more likely to detect patterns that are not simply attributable to chance). However, sample sizes are generally constrained by practicalities, economics and ethics. In the absence of any other guiding factors let us decide on 10 data points per sample. You can look up how bone density is typically measured, and typical values you might expect, in the literature or online. These sources should indicate that bone density is typically measured in grams per cubic centimetre, that values between say 0.6 and 1.1 would be realistic, and that values are typically quoted to two or three decimal places. From this you might come up with the following dataset:

DUMMY DATA Bone density measurement ($g\ cm^{-3}$)	
USA	**China**
0.72	0.80
0.92	1.01
0.84	0.87
0.93	0.74
1.04	0.86
0.75	0.81
0.79	0.79
1.07	0.96
0.86	0.65
0.75	0.72

7.4.4   Worked example 2   Choosing a statistical test

Use the dummy data from Worked example 1 to answer the following questions; then decide which statistical test you would use to determine whether bone density differs in adult females over 50 years of age in the USA compared to China. Note that parametric tests make assumptions about data that non-parametric tests do not (see Hawkins (2009) for further information).

Question 1: what do I want to do?

- *Option 1.1* Perform tests on frequencies (i.e. the distribution of one or more sets of observations), where the data are frequencies or counts in categories. **Go to Question 2**.

- *Option 1.2* Test the relationship between variables, where the data are scale (measurements or counts) or ordinal (ranks). **Go to Question 3**.

- *Option 1.3* Test the difference between sets of observations, where the dependent data are scale or ordinal, and in categories defined by the independent variable. **Go to Question 4**.

Question 2: how many categories are there?

- *Option 2.1* There is one set of categories, so a one-way classification including a test of homogeneity is appropriate: **one-way Chi-square test.**

- *Option 2.2* There are two sets of categories, so a two-way test is appropriate (also called a *test of association* or a *test of independence*): **two-way Chi-square test**.

Question 3: is there a dependent variable and an independent variable?

- *Option 3.1* Yes, there are clearly identified dependent and independent variables, so that a regression is appropriate.

  ○ Either, the data are parametric: **linear regression**.

  ○ Or, the data are non-parametric: **non-parametric regression**.

- *Option 3.2* No, the variables are interdependent and a test of correlation is appropriate.

  ○ Either, the data are parametric: **Pearson correlation**.

  ○ Or, the data are non-parametric: **Spearman correlation**.

Question 4: are samples unrelated or related?

- *Option 4.1* The samples are unrelated. **Go to Question 5**.

- *Option 4.2* The samples are related. **Go to Question 6**.

Question 5: are there two samples or more than two samples?

- *Option 5.1* There are two samples.

- ○ Either, the data are parametric: **t-test.**
- ○ Or, the data are non-parametric: **Mann–Whitney U test**.

- *Option 5.2* There are more than two samples.

  - ○ Either, the data are parametric: **one-way ANOVA**.
  - ○ Or, the data are non-parametric: **Kruskal–Wallis test**.

Question 6: are there two samples or more than two samples?

- *Option 6.1* There are two samples.

  - ○ Either, the data are parametric: **paired t-test.**
  - ○ Or, the data are non-parametric: **Wilcoxon signed-rank test**.

- *Option 6.2* There are more than two samples.

  - ○ Either, the data are parametric: **related one-way ANOVA**.
  - ○ Or, the data are non-parametric: **Friedman ANOVA.**

The correct answers are indicated at the foot of the page[*].

7.4.5 Finding and evaluating resources

You should now be at the point where you have selected a particular technique, or at least a limited range of techniques. What you are likely to need at this stage is some help with carrying out these technique(s). There was a time when such help was scarce, particularly help designed especially for biologists. However, there is now plenty of support available to you. This includes a range of books on statistics for biologists (see Hawkins (2009), and selected further reading therein); books designed to support biologists in other areas of maths are starting to emerge, for example, Reed (2011). Of course, online resources have also been developing rapidly over recent years, and the relevant websites are listed in Section 7.8, Additional resources, at the end of this chapter.

Your greatest challenge is likely to be finding good, relevant resources quickly and efficiently. The *SUMS-finder* website is a tool designed to help with this (see Additional resources). The site has a facility that allows resources to be searched by resource type (interactive object such as an animated activity; a web page with navigation and different media; and/or a printable document), accessibility (for example you can exclude sites that require subscription) and level ('basic', 'intermediate' or 'advanced').

Another good way of finding resources is to ask around for personal recommendations. You need to overcome any social stigma you may feel about talking about maths and stats to your peers – remember My Maths Mantra (Section 7.3.1: 'I won't lose all my friends')!

---

[*]**Answer:** Question 1: Option 1.3 → Question 4: Option 4.1 → Question 5: Option 5.1 – either a t-test or a Mann–Whitney U test depending on whether the parametric criteria are met.

7.4.6 Tutor notes   Helping students acquire skills

There is no substitute for *doing* when it comes to maths and stats: lectures have limited value. Whenever possible, teach through tutorials rather than lectures. If and when you do give a lecture, make it as interactive as possible.

It is important to focus on the quality of what students learn and not the quantity. If a tutorial is not going so well, far better to make sure students understand half of the material well rather than not really understanding anything. This approach should help build confidence.

In addition to enthusiasm (Section 7.2.4), those teaching maths and stats to biologists need to have patience. Students of the biosciences need you to understand that sometimes it may take them some time to acquire a skill. With the more difficult problems, students need to be encouraged to think about things more, ideally puzzling over them in a fun way. Again, taking sufficient time to do this well is important.

Finally, being able to find good, relevant teaching materials may be a challenge. The *SUMS-finder* site discussed in Section 7.4.5 will be helpful to staff as well as students.

7.5   How you can achieve your potential in biomaths

- Understand how very useful maths is as a tool for a biologist.

- Become competent at the basics.

- Develop overviews of more advanced material.

- Look for commonalities which reduce the amount that you have to learn.

- Be patient, focus on quality of learning not quantity.

- Remember the biomaths mantra: you are not alone – you don't have to become a mathematician – getting the wrong answer doesn't mean that you don't understand – you won't lose friends – you can do it!

- Be proactive in finding resources to help you.

- Have fun!

7.6   Acknowledgements

Thanks to David Adams, Rachel Camina, Toby Carter, Jenny Koenig, Julian Priddle, Vicki Tariq and Mike Weale for useful comments on, and corrections to, the draft of this chapter.

7.7   References

Carter, T., Priddle, J., Hawkins, D. and McCary, J. (2010) SUMS (Students Upgrading Mathematics Skills). *SUMS project final report version 5.2* to HEA/EvidenceNet. August 2010. Available to download free from www.anglia.ac.uk/tobycarter. Also see www.step-up-to-science.com/SUMS/. Accessed April 2011.

Cohen, J.E. (2004) Mathematics is biology's next microscope, only better; biology is mathematics' next physics, only better. *PLoS Biology* **2**, 2017–2023.

Hawkins, D. (2009) *Biomeasurement* (2nd Edition). Oxford: Oxford University Press.

Hardy, G.H. (1908) Mendelian proportions in a mixed population. *Science* **28**, 49–50.

Jungck, J.R. (1997) Ten equations that changed biology: mathematics in problem-solving biology curricula. *Bioscience* **23**, 11–36.

Lee, S., Browne, R., Dudzic, S. and Stripp, C. (2010) *Understanding the UK Mathematics Curriculum Pre-Higher Education*. Loughborough: The Higher Education Academy Engineering Subject Centre. Available for free download at: www.bioscience.heacademy.ac.uk/ftp/resources/pre-university-maths-guide.pdf. Accessed April 2011.

May, R.M. (2004) Uses and abuses of mathematics in biology. *Science* **303**, 790–793.

National Mathematics Advisory Panel (2008) *Foundations for Success: The Final Report of the National Mathematics Advisory Panel*. Jessup, MD: US Department of Education.

Reed, M. (2011) *Core Maths for Biosciences*. Oxford: Oxford University Press.

Ruxton, G.D., and Colegrave, N. (2011) *Experimental Design for the Life Sciences* (3rd Edition). Oxford: Oxford University Press.

Tariq, V.N., Durrani, N., Lloyd-Jones, R. *et al.* (2010) *Every Student Counts: Promoting Numeracy and Enhancing Employability*. Preston: University of Central Lancashire. Available for free download at: www.uclan.ac.uk/information/services/ldu/files/ESC_FINAL.pdf. Accessed April 2011.

Thomas, A. (2003) *Introducing Genetics: From Mendel to Molecule*. Cheltenham: Nelson Thornes.

7.8 Additional resources (accessed April 2011)

Biomeasurement 2e online resource centre. Material supporting Hawkins (2009), specific to introducing statistics to bioscience students. www.oup.com/uk/orc/bin/9780199219995/.

GeoGebra. Free software for learning and teaching maths, supported by an international community of users; potentially of great use to biologists. www.GeoGebra.org.

getstats. The Royal Statistical Society's campaign to 'build a society in which our lives and choices are enriched by understanding statistics'. www.getstats.org.uk.

mathcentre. Developed by a group from the universities of Loughborough, Leeds and Coventry to provide free mathematics material at the post-16 level, including support specific to the biosciences. www.mathcentre.ac.uk/.

Biomathutor. Represents a prototype multimedia e-learning resource which aims to support mathematics learning in the biosciences. www.ebst.co.uk/biomaths/intro.html.

NRICH. Based at the University of Cambridge, this free resource is particularly directed at school level and has a section dedicated to STEM subjects, including biology. http://nrich.maths.org/.

NUMBAT. A free resource supporting numeracy and statistics for students in Further and Higher Education, developed by a group at Anglia Ruskin University. www.numeracy-bank.net/.

NuMBerS. The Numerical Methods for Bioscience Students project was designed to underpin the teaching of numerical methods for the biosciences, with a range of free information sheets. www.anglia.ac.uk/numbers. This material is also available through the NUMBAT site.

Science ETC. A consultancy delivering education and training in quantitative bioscience, especially pharmacy and biochemistry, including some free e-learning resources. www.sci-etc.co.uk/.

SUMS-finder. A portal under development for searching and evaluating resources designed to support maths and stats across the curriculum in Further and Higher Education. www.step-up-to-science.com/sumsv3/.

8 E-learning for biologists

Jo L. Badge, Jon J. A. Scott and Terry J. McAndrew

8.1 Introduction

Whatever your intended career path on graduation as a life scientist, it is highly likely that technology will play a significant role in your day-to-day work. Online, remote working is becoming increasingly commonplace, and this may involve facilitation of international collaborations that are reliant on video conferencing, online data storage and retrieval, and the use of other network tools. A great wealth of information is available online, and graduates need to be skilled in the methods of gathering and filtering information as well as critically appraising it. A recent e-skills report entitled 'Technology counts' estimated that 77% of UK jobs require information and communications technology (ICT) competence, with skills that need to be frequently updated (eSkills UK, 2008).

You will need, therefore, to acquire the necessary *digital learning literacy*; that is, the appropriate ICT learning skills and knowledge that will enable you to be an effective student of the modern biosciences. You may believe that these skills are embedded in your courses and that you will become well versed in all the appropriate technologies – but this is only part of the story. The digital environment is always changing rapidly, which is why in this chapter we talk about the broader ideas, with some current examples to illustrate the typical skills useful to a student in the biosciences.

There are numerous tools available online (many of them free) to assist your studies and facilitate your move to employment or further study on completion of your degree. This chapter aims to help you to be an effective learner online. It is not only about the tools you could use but also about how to use them.

8.2 Online working environment

Any bioscience degree programme involves a learning path. During your journey along this path, you will develop a variety of skills alongside the information framework for your subject area. The development of digital skills will play a vital role in the effectiveness of your overall academic study; for example, these skills will enable you to search for and access information. It is, therefore, helpful to decide where you are on the path to developing digital competence so you can ensure that you continue to move forward.

Initially you need to assess your current skills level and identify any deficits you may have with respect to your course requirements. There are, however, many ICT skills that could be useful and it

Effective Learning in the Life Sciences: How Students Can Achieve Their Full Potential, First Edition.
Edited by David J. Adams.
© 2011 John Wiley & Sons, Ltd. Published 2011 by John Wiley & Sons, Ltd.

can be very difficult to know which ones you need to pay attention to, and in what order. Since most universities adopt an online/web-based approach for delivery of their courses, a good starting point is to become familiar with the online facilities your institution provides.

8.2.1 The 'portal' and the 'virtual learning environment'

In recent years, universities have adopted two new terms to describe the online support systems they provide for their students. The first of these is a *portal*, which acts as a doorway to all of the web resources laid on for students. The portal often contains general administrative information including course guides with timetables. Some universities have equipped their portal with a mobile interface so that you can gather information whenever you need it. Unfortunately this facility can encourage last-minute changes, and you may miss important announcements. Watch out for this: keep up to date with what is happening and always plan your workload well in advance.

The second term/facility you are likely to encounter soon after arriving at university is the *virtual learning environment* or VLE. The portal will provide a direct link to the VLE. The VLE will contain a great deal of online information relevant to your course, for example lecture notes, laboratory schedules, sites for electronic submission of coursework and links to a wide range of additional resources.

It is likely you will feel a little overwhelmed by the amount of online information available to you when you begin at university. Your tutors will help you navigate a route through all this information, so listen to their advice and take notes. The VLE will probably reveal information about your course/modules in a time-controlled manner; that is, material will be released to you as necessary, on a rolling basis. This should mean you are guided to the appropriate material at the appropriate time.

The VLE will usually offer discussion boards which enable both students and staff to post contributions on topics relevant to the course. These can be really useful as they are dynamic and have a reasonably short response time: staff can monitor comments made by students and post responses to questions about course content for all to see. Many common concerns can be dealt with rapidly and efficiently in this way. Discussion boards are technically very easy to use once you have overcome the fear of sharing your thoughts online. Build confidence in using these systems by participating in online discussions as soon as you get the opportunity.

The VLE should contain all the information about your modules that you will need. But you may also be given a hard copy of the course guide which often contains the answers to frequently asked questions; you can save yourself some time online by reading it carefully. You should find out how to search the VLE for other useful information (search systems should be made plain on screen – if in doubt ask an administrator). You should, for example, be able to access information about modules available in future years, and this may help inform your choice of modules at Levels 2 and 3 of your programme.

8.2.2 Tutor notes

Many students find an introductory course guide (provided in hard copy and/or made available online), with a set of *frequently asked questions*, a valuable supportive resource to allow them to become quickly acquainted with the university's specific systems and their operation. Make good use of the automatic date settings on your VLE so that you can feed information to the students at appropriate times in the module. These date settings can also be used to hide information that is no longer required; therefore automatically keeping the VLE relevant and timely.

8.2.3   Making the most of tools and technologies accessible from the VLE

Your university will ensure that you have access to a wide range of tools; for example, literature search engines such as *PubMed* and *Web of Science*. You may not enjoy this level of access to these tools after you graduate, so make the most of this opportunity to become familiar with cutting-edge technologies, and develop skills that will prove invaluable during your studies and after you graduate from university.

8.2.4   Your personal learning environment

As your studies progress, you will build a collection of online and other resources that you use frequently. This will include the software available on your personal computer, tools and technologies provided by your university, websites of publishers, freely available public resources etc. This collection is often described as your *personal learning environment* (PLE). Your PLE will be dynamic: you will acquire and delete elements on a regular basis. New websites and applications, in particular, appear on a daily basis and you will find it impossible to evaluate all of these. This is where social networks can be very valuable, as you benefit from the experiences of others.

8.2.5   Your social network – offline and online – keeping the balance

Most students belong to a social network of some type long before they get to university. These networks offer a very easy way to keep in touch with friends and with what is going on. When you arrive at university you will have the opportunity to markedly extend your network as you make new friends and acquaintances. You will also be able to use the network to share, with fellow students, experiences and resources of relevance to your course, and to collaborate on academic work. This will provide a new challenge, as you may now find yourself using the same resource both for socialising and for work. You will need to strike a balance between the two: make sure you regulate the amount of time you set aside for online socialising, as it is very easy to be distracted from work. When you are working, you can reduce the number of interruptions by setting social network privacy controls to limit the audience to those who need to see your postings. If you are using email, you can switch off notifications advising of incoming mail. You can limit access to Facebook to set times, and deal with all of the incoming traffic at once. By all means have fun with social networking tools, but try to keep work and play separate (and don't be distracted by online games!).

8.2.6   Keeping up

In recent years we have seen a boom in 'Web 2.0' technologies. These are mostly encountered in the social networks and services e.g. Facebook, YouTube, Flickr, Blogger etc. Web 2.0 technologies can also have an academic utility, for example through social bookmarking services such as Delicious or online reference managers such as CiteULike or Mendeley. These are rapidly evolving areas, and the resources can help you become a more effective student. However, constantly striving to keep up to date may not always be the best use of your time. Try to be selective and use a relatively small number of tools. You can always review your use of them once or twice a year to see which are working well and which may have been superseded.

8.2.7 Tutor notes

Many members of staff may have anxieties about their ability to engage with all of the resources available online. It may be useful to organise staff development sessions for colleagues to provide an overview of the online tools that you are using in your module(s). Consider introducing only one new service at a time and evaluate its success before introducing additional online tools. The HEA Centre for Bioscience provides some excellent links to e-learning resources that are regularly updated and contain real-life examples of effective practice in higher education (www.bioscience.heacademy.ac.uk/resources/themes/elearn/elearncs.aspx).

8.3 Resources

8.3.1 Mobile phones

Universities are likely to make use of technologies they consider readily accessible by all of their students, and mobile phones are becoming increasingly important in this regard. For example, so many students now have mobile phones capable of receiving Short Message Service (SMS) messages that SMS may be used to issue reminders or warnings about last-minute changes to course events. If your phone has sound-recording facilities or a camera, you might find it valuable to use it, for example, to record audio notes or to take a photograph of an experimental setup during a laboratory class; the photograph could then be used as part of the write-up for the experiment. However, a word of caution concerning the use of mobile phones in laboratories: there may be safety considerations relating to possible contamination of the phone by, for example, chemicals or microorganisms; so make sure you are aware of any local regulations regarding the use of these mobile devices. Of course, you could use the audio recording facility on your phone (or other 'Dictaphone' device) to record lectures or tutorial sessions and review these recordings later, at your leisure. However, make sure that you first ask permission from your lecturer or tutor. Please make sure you don't use such recordings as an excuse for not taking notes during teaching sessions, as listening again to entire lectures is very time consuming. Rather, simply use the recordings to help you fill in any gaps in your notes.

8.3.1.1 QR codes and podcasts on your phone

Some mobile phone manufacturers, including Android, Nokia and Sony, offer a two-dimensional (2D) barcode reader as an application. Apps (small software packages) for these can be downloaded from the appropriate online support store. The most popular type of 2D code is the QR code format (http://reader.kaywa.com/; Figure 8.1). These systems link the physical world with the virtual one. For example, a display poster about a specific topic could be linked to an online discussion board in the VLE that contains further information and gives you and other students the opportunity to make comments on the content of the poster.

Universities and other providers are making more and more podcasts of lectures and other presentations available to students or wider audiences. These can be accessed using mobile devices such as smartphones or mp3/mp4 players, but we have noted that they tend to be viewed, or listened

Figure 8.1 Example of QR code. This code will direct you to the BBC feed for science and environment news, http://feeds.bbci.co.uk/news/science_and_environment/rss.xml

to, on broadband networks at home or in halls where students have ready access to their notes and writing materials. Podcasts can be convenient and valuable learning tools, but try to limit your use of these resources as there is clearly a limit to the number of podcasts you can view and absorb in a typical week.

8.3.2 Tutor notes *Cameras and other recording devices associated with mobile phones*

It is advisable to have a clear departmental policy regarding the use or otherwise of mobile phones in different teaching contexts. For example, with regard to the cameras built in to these devices, while it may be acceptable for students to photograph the equipment setup in a lab prior to experimental work, it would clearly be inappropriate for them to do so whilst handling live specimens, chemicals etc. Students should also be informed that they must seek the permission of individual students or members of staff before taking their photographs.

Some individuals, for example dyslexic or international students, may benefit greatly from making audio or video recordings of lectures for personal use. Again, clear guidance regarding the making and use of recordings is important. In particular it should be made very clear to students that they should not post any photographs or sound recordings of teaching sessions on external websites.

8.3.3 *Your personal computer vs. university computers*

Your laptop may have limited application on campus, as most provision is geared to software installed on the computers provided in university buildings. The computers the university provides on campus will be optimally configured for student use, very well supported with technical help and know-how, and furnished with specialist software for class activities and practical work; e.g. software that simulates complex laboratory procedures or that facilitates statistical analyses (see Section 8.3.5, Choosing appropriate software for your personal computer). However, while in the main university buildings on campus, and often in halls of residence, you will be able to use your laptop to read emails, log on to the VLE and obtain at least limited access to university software, and this will normally be through a wireless network. This 'remote access' facility should also enable you to access university printers from your laptop, a facility that is particularly useful when you wish to print out material during a visit to the library or other communal area. You will need to find out about the university's remote access requirements (username, password etc.) before you can use the wireless facilities.

8.3.4   University printers

Most universities provide and sell printer credits that enable you to print a specified number of pages on university printers. Take some time to think about whether you really need to print materials, and save time, money and the environment. When printing slides from lectures make sure you choose the 'n slides per page' option to print multiple slides per page. Each printer has 'properties' and 'options' which can be changed, and the more sophisticated high-volume printers can typically be set to print more than one page on a sheet of paper: you could reduce your printing bill by half!

8.3.5   Choosing appropriate software for your personal computer

Students are often attracted to the most basic software simply because everyone else seems to be installing it on their computers. They feel confident that if they have problems then there ought to be lots of experienced people around who can help out. Do cast your net more widely than this, because if you choose the most popular, basic software you may miss out on more sophisticated software that can potentially do a lot of work for you. For example, Microsoft Excel is often recommended for basic statistical analysis because it is software that graduates are likely to encounter when they leave university. However, you may find that software capable of more sophisticated statistical analysis is more appropriate for your subject and course, for example Minitab, SPSS, R or Stata (for more information see Section 8.12, Additional resources). Universities hold site-licences for many of the software applications used on campus. Essentially they are in a position to buy in bulk and may therefore be able to offer you software, including some specialist packages, at low cost. If you want to try out software, a company will often agree to a trial period. If you are impressed you can then access the software, for a small fee, through the university's site license.

   You may also wish to consider using free and open-source versions of popular office programmes. OpenOffice (www.openoffice.org) is an excellent and free alternative to Microsoft Office that includes a word processor, spreadsheet, and presentation and drawing software. It will run on Windows, Mac or Linux computers. There are also free open-source versions of email clients (e.g. Thunderbird). It is possible to load all of your favourite open-source applications onto a USB stick and run them directly without having to install them more permanently on your computer. This allows you to take the software with you wherever you are working and therefore replicate your working environment on different computers (see http://portableapps.com/).

The choice of web browser available on your university's computers will be limited to versions compatible with the institution's VLE and other web services. You should follow any recommendations provided by the institution, but be aware that there are many browsers available besides Microsoft's Internet Explorer that is found on most institutional PCs: try Mozilla Firefox, Opera, or Google Chrome as alternatives (see Additional resources). Chrome and Firefox each use a system which allows for tools to be installed in your browser; these extensions enable all kinds of features, for example automatic notifications for online calendars, email, Facebook or Twitter.

Online office suites are available from a number of sources. You don't need to provide software to use these suites; they will allow you to work on your documents at any site with Internet access. They include: Google Documents (www.docs.google.com), Zoho (www.zoho.com), ThinkFree (www.Thinkfree.com) and Adobe Acrobat (www.acrobat.com). All of these offer full word-processing software; you can upload existing documents in a wide variety of formats or create new ones online.

Online office suites offer the additional advantage of online collaboration in real time. If you and your fellow students need to complete a piece of group work, you can create a single document online and give permission to members of the group to edit it. A record of the contributions made by each of you will be tracked in an open revision history. This makes editing the document much simpler because there is always only a single agreed version of the document, rather than multiple versions sent round by email. You can work at a time and place that is convenient to each of you: from home, in the library or, during vacations, at sites that may be hundreds of miles apart.

8.3.6 Accessibility and special needs in the use of computers

You may have special requirements regarding access to e-learning resources; if so, your university should have procedures in place to support your needs. Sometimes, you may not discover an accessibility issue until you reach university and leave the supportive environment of your school; perhaps because you have not previously used online resources. The on-campus disability support staff should work with the information and communications technology (ICT) providers at your university to make sure that adequate support or alternative software is in place. However, you can also find out what is available at www.jisctechdis.ac.uk/accessibilityessentials. If you have any problems working with courseware, contact your tutor immediately and discuss this so that all learning resources can be adjusted as appropriate.

8.3.7 Tutor notes

Improved accessibility of teaching materials should not be an onerous burden for the academic preparing resources for teaching. Good-practice guidance and support ought to be available within the institution, and can be accessed on the Internet, e.g. JISC Techdis (www.jisctechdis.ac. uk/). The enhanced accessibility of materials is also likely to benefit international students. For example, ensuring that audio or video files have accompanying text transcripts, primarily for the benefit of students with hearing problems, will also benefit students with English as a second language. A good place to start is the JISC Techdis Accessibility Essentials series, which will show you how to make electronic documents more readable, and Powerpoint presentations more accessible (www.jisctechdis.ac.uk/accessibilityessentials).

8.3.8 Case study EduApps

Figure 8.2 EduApps USB stick

EduApps is a collection of open-source applications that can be downloaded free or accessed from a USB stick (Figure 8.2). The initiative improves access to a range of technologies. There are several packages:

- **AccessApps** Supports writing, reading and planning, and deals with sensory, cognitive and physical difficulties.

- **TeachApps** A collection of software specifically designed for teachers or lecturers.

- **LearnApps** Specifically designed for learners. Includes 'sticky notes' for reminders and annotating documents, a free word processor, 'Audacity' for sound recording and editing, universal media player, dictionary, 'KeePass' password manager, periodic table.

- **MyStudyBar** Provides a suite of apps to support literacy, such as 'Orato' text reader, mind mapping tool, 'Let me type' (word prediction), talking dictionary.

- **MyVisBar** A high-contrast, floating toolbar designed to support learners with visual difficulties.

- **MyAccess** A way to have your favourite EduApps in one place on your PC, plus a range of tools to ensure e-learning materials are accessible to all.

- **Accessible Formatting WordBar** Enables creation of accessible Word documents containing text that can be read aloud by screen readers; invaluable for visually impaired or dyslexic students.

See www.rsc-ne-scotland.ac.uk/eduapps/ to download EduApps or get more information.

   Acknowledgement: case study reproduced by kind permission of JISC Regional Support Centre Scotland North & East.

8.3.9 Finding and using great images

You will often need to find images to illustrate your coursework, or to help you better understand a particular topic you are studying. There are numerous places you can look online for good quality images that are free to use. Check the information in recommended textbooks, as many publishers provide free online study guides to accompany their books. These often contain images used in the books or supplementary information (note that you may need to buy the book first in order to obtain a login key to access the materials).

Good sources of images online are photographic sharing sites such as flickr (www.flickr.com) or picasa (picasa.google.com); users can search for photos online then download the images to their computers. Special biological image databases are particularly useful for images of organisms; for example the Higher Education Academy UK Centre for Bioscience hosts *ImageBank* (www.bioscience.heacademy.ac.uk/imagebank/), which is free to use for educational purposes. Further image banks and other resources specifically for biologists can be found at www.academicinfo.net/bioimage.html. However, not all images accessible on the Internet are freely available and you must take care not to breach copyright when copying images for your own use.

8.4 Legal considerations

Copyright is easily abused by accident. You need to be aware of your duties and responsibilities with respect to copyright of the work of others. This can be a very difficult legal area, but the following should help.

(1) Remember that if someone posts their work on a public site such as YouTube, this does not automatically grant you rights to include the material in your own website or blogs.

(2) Look for material granted 'Open' rights; this means that others are free to use it without restriction. A safe way to do this is to search for material licensed through the Creative Commons initiative (http://creativecommons.org/). In addition, the advance search facility on Flickr will identify images that are licensed for re-use.

Remember that material you post online is likely to be around for a very long time; so be mindful that risks associated with the use of content without permission are not necessarily short term. Think carefully about what you write online, in public spaces, about others, including your lecturers and university. You could find that these public statements are not interpreted by others in the way you intended (see also Section 8.7, Developing as a professional).

Finally, as with all sources of information, you must make sure that you refer to online material correctly during any subsequent use or publication. This issue is addressed in Section 8.9, 'Working effectively'.

8.4.1 Tutor notes

It is important that lecturers set a good example, with respect to the use of images and other online materials, during teaching. They must ensure that they comply with copyright and that all images and other resources are attributed appropriately.

8.5   Protecting your work

Electronic copies of coursework are easily lost, and you should do all you can to protect your property. There are at least three major threats to the security of your work.

(1) **Failure to make backup copies:** each year students lose electronic copies of significant items of coursework because they made only a single copy that gets damaged or misplaced; losing significant marks due to such a simple error is very upsetting. Make sure you always make a backup copy of your work and save it at a location or on another device remote from your computer. Memory sticks can be used as backup devices but, of course, these can also be lost. A better bet is to make use of the university's computing system. The computing 'account' your university provides will include file space that you can readily access through the network and that the university will backup for you. This is a very good place to store a second copy of an important document.

(2) **Accidents and viruses:** you may accidentally overwrite your work with other work (technically known as 'clobbering'), or your work may be corrupted through computer virus infection. Keep filenames meaningful to the work and include dates or version names (for example 'evolution_ essay_12_05_10' or 'evolution_essay_v1') to avoid overwriting with earlier versions. Avoid viruses by installing antivirus software provided by your university. Most institutions will have a site-licence policy that includes use by students. While you are registered as a student you can install the antivirus software on your own laptop and avoid paying annual licence fees.

(3) **Malicious damage by others:** if you are lax with security, others have the opportunity to steal your work, and present it as their own, or simply to make a mess of your file space. You need to be very careful with the passwords to your accounts.

8.5.1   Working with passwords

Nearly all online services will require you to register and create an account with the provider. This can lead to lots of usernames and passwords that you must recall. Passwords are important: you are likely to store coursework or personal information online and it is essential that this material is protected. Passwords need to be secure and you need to be able to remember them! There are two strategies that can help: the first is to use an online password storage facility and the second is to develop a password mnemonic phrase system.

8.5.1.1   Utilising password storage systems

These systems allow you to use one password and username to access a site that contains all of your other passwords and usernames. This site could be located on your computer or accessible online. A word of warning: we would not recommend these systems for storing password details of anything that must be kept very secure, such as bank details, access to online banking or credit cards! You should always try to commit these details to memory. However, password storage systems can be very useful for freely accessible tools, such as those described earlier in this chapter, or services you simply wish to explore. Examples of password storage systems are:

Online systems
www.Clipperz.com: this site can create randomly generated passwords and can be used to store them. It can provide direct links to websites by logging in to these sites and opening the service in a new tab or browser window.

Single sign on

Many services offer the ability to sign on via other accounts you may already hold: these include Yahoo email, Gmail (Google Mail), Twitter or Facebook. Whilst this is often a quick and easy option, think carefully about linking systems you wish to keep secure to your social networking sites, and remember to carefully manage the third-party access you may allow via these accounts.

8.5.1.2   Mnemonic password phrases

A good mnemonic password phrase will allow you to remember a password easily and avoid having to write it down. A good, strong password contains letters, a mix of cases and numbers and, if possible, non-alphanumeric symbols (such as £@$%*). An easy way to incorporate all of these in a single password is to develop a phrase that can be adapted for each new password you need. The phrase should be unique to you. For example, a password for a Google account could be made as follows:

John **c**an **p**ass **h**is **G**oogle **e**xam **w**ith **f**lying **c**olours **i**n **t**wo **d**ays

Take the initial letters of this phrase and substitute numbers where appropriate:

jcphgewfci2d

Use some upper case letters:

JcphGewfci2d

Replace the i with an exclamation mark (!)

JcphGewfc!2d

Each time you need a new password, replace the word 'Google' with the system you are using; for example, a password for Adobe.com could be: jcphAewfc!2d where 'A' is substituted for Adobe in the phrase. One advantage of a phrase is that you can 'say' the phrase to yourself while you are typing to help you remember the resulting password.

8.6   Organisation

'Failing to plan is planning to fail' goes the saying, and while you may be able to survive with little planning, you will not thrive unless you can manage your time effectively.

8.6.1   Using online calendars

A good diary will help you manage your time. The diary does not have to be electronic but computer-based systems certainly have advantages. Google Sync (www.google.com/mobile/sync/) is a good example of a service that will coordinate the calendar, contacts and email facilities on your mobile phone and synchronise them with other points of access to calendar, contacts and email such as your laptop or a web browser. Other applications, e.g. Mail for Exchange, can work with your inbox on campus. These applications will automatically update any changes across all platforms. You can also set reminders for important coursework deadlines. Google Calendar allows you to send reminders to your phone as text messages.

8.6.2   Using 'to do' lists

Some people find a 'to do' list a great way to stay organised; for others it can present an overwhelming collection of work in need of completion. If you like to keep lists, consider doing so online so you can work from one consolidated list in an organised fashion. For example, Remember the Milk (rememberthemilk.com) works on a wide variety of mobile platforms and will send reminders to you by text and automatically enter items in your online calendar. Email software, such as Microsoft Outlook or Thunderbird, also often includes task managers.

8.7   Developing as a professional

During your time at university you should think carefully about how you are portrayed online and develop a professional identity, as this will enable you to graduate with a track record of achievements to show prospective employers.

8.7.1   Establishing your online reputation

Social networking online is a way to connect with friends all over the world regardless of their physical location. Networking sites such as Facebook will probably play a large part in enabling your social connections with friends at university, at home and those from school who are now at other universities. Recent concerns about privacy on Facebook (Hoadley *et al.*, 2010) have highlighted the need for users to think carefully about how their accounts are set up and managed. A further cause for concern is that, increasingly, employers are looking at job applicants' online credentials and conduct as part of their recruitment processes. Even in academia, your name may be 'Googled' when you apply for a place on a Masters course or a PhD position. Developing a professional and responsible persona online may therefore be essential for your future career.

Exercise

Take a moment to check how you appear online now: type your name into an online search engine, for example Google, and see what comes up. Remember to check different spellings of your name (for example, Jo, Joanne, Joanna), or try a username or any pseudonyms you use online. You may be surprised by what this generates, and if you can easily locate this material so can anyone else. This should remind you to think carefully about who has access to your material; don't forget that people may be able to access information about you indirectly, for example through the pages your friends have set up.

Effectively managing your identity online is clearly essential, and for the first time you may need to begin to think about having separate private and professional identities. You will have a university email account and no doubt you will already have at least one private email account. You could just as easily create separate accounts for other online tools and systems you may want to use. If your Facebook account contains personal information, or photographs of you socialising that you would prefer a potential employer or your tutor not to see, then make sure that privacy settings are managed well so that privileged information is shared only with approved friends and is not open to public searches. Remember that material on the Internet is effectively there forever; deleting a Facebook account is not simple, so regularly check your presence online by doing what any employer might and 'Google' yourself!

Developing a presence online that is professional and well managed may benefit your future prospects in other ways. For example, microblogging sites such as Twitter.com could provide a place to network in a professional capacity, or perhaps you could set up a study group with your fellow students? If you have established a trustworthy reputation online, this will help you work effectively online with others.

As you build an online identity you will use online tools to your advantage and share the knowledge you acquire with others. Much has been made of the Web 2.0 revolution, where content is created by users rather than provided by authoritative sources. Many of the Web 2.0 tools, and in particular social networking tools, are only as good as the network you create around them. The content and networks you create, and shape in collaboration with others, will therefore affect the reputation you build online.

8.7.2 Tutor notes

Tutors need to respect the boundaries between public and private social networking spaces as much as students need to learn to differentiate between the two. Students generally do not welcome the intrusion of staff into private networking spaces such as Facebook. So, while it may be appropriate to create specific pages on Facebook for information relating to teaching which does not require staff to make online connections or 'friendships' with students, it would be inappropriate to exchange 'friend requests' with students. Staff may find it helpful to reinforce these professional and private boundaries by having professional online identities on social networks that speak to a different audience and are separate from their private identities. For example, an 'academic' Twitter account versus a private, family Facebook account.

8.7.3 Developing a portfolio

One way to establish an effective online identity is to create a portfolio of your work, progress and achievements in your own space online. It is also likely that you will be asked by your tutors to keep a portfolio of work for personal development planning (PDP). A simple way to fulfil both of these requirements is to write a blog. You can keep this very simple to start with, using the blog as a place to collect your thoughts on your studies, and it can act as a central hub for collation of the other sources of information you find online. A simple blogging system such as Posterous.com will enable you to very quickly create content by simply emailing photos, videos, bookmarks or text to your account and this will then be published online. The blog will provide online space that is within your control and will allow you to project the professional image you may need to impress potential employers.

8.8 Information online

At times you may feel overwhelmed by the amount of information available online. The identification of material relevant to your studies can be time consuming, and it is important to check the academic quality of this online information. Here we examine strategies that should help you cope with the tide of information flowing over you.

8.8.1   Finding valid sources of scientific information

It is essential that you check the validity of the information you access. Most scientists use peer-reviewed papers as their main source of information. Peer review is a system by which two or more scientists assess the quality of written reports of research compiled by other scientists. If the work is considered of sufficiently high standard then it is published in a reputable scientific journal. You can be very confident about the quality and validity of information presented to you in a paper published in a high-quality journal like *Nature*. But what if you just want to quickly check a point of information online? Most of us will turn to an online search engine for the answer, and it is likely that the top hit will be an article in 'Wikipedia'. Once scorned by academics as an unreliable source of information, Wikipedia now contains many very useful entries that often lead to more reliable sources of information. However, anyone can edit a Wikipedia page and it is therefore extremely difficult to ascertain the validity of all of the information presented. When you read an article at the Wikipedia site, you should therefore be critical and look carefully at the literature cited in the article – does the article contain references to peer-reviewed journal articles, and are these listed at the end? Does it have any Wikipedia warning notices that the information is incomplete, controversial or under review? You should back-up any initial research you do, based on information provided at the Wikipedia site, with information from other sources including lecture notes, textbooks recommended for your course and peer-reviewed articles in reputable journals (ask your tutor if you are in doubt). The Wikipedia site should never be given as a formal reference source, in its own right, in the list you compile at the end of an essay or other form of coursework.

When you are undertaking a comprehensive search of the published literature you should be aware that the highest ranking result from an online search will not necessarily be the best source of information. If you found the source easily in this way, it is likely that your fellow students will do the same. Your work is therefore not likely to stand out from the work of others and you may not obtain a high mark for the assignment. You need to make more of an effort and take more of an academic approach during a search and you should bear in mind the following.

- Your library will provide you with access to many academic journals online (see Section 8.8.3). You can therefore search through original material directly and easily, from your desktop computer, without need of a Google or other search.

- Information specialists, like librarians, will be happy to guide you to resources and teach you how to obtain the best results from searches. They are friendly, approachable people with many years of experience, and working with them will save you a lot of time. Think of a knowledgeable friendly expert as the ultimate piece of software – they will actually understand the questions you want to ask!

- The library/study skill centre at your university is likely to run short courses on information skills, involving databases and search engines, so sign up for a course at an early stage in your degree programme.

- Make use of online tutorials that will also equip you with good searching skills and suggest strategies you can adopt. These include the subject-specific Virtual Tutorials provided by Intute (see the Virtual Training Suite case study).

You should make the most of the various specialist, academic search engines and databases provided by your university. These are likely to include Web of Knowledge, Biosis Previews and Medline, and they act as gateways to peer-reviewed papers and articles that you will find most useful during the later stages of your university career. Finally, you may find 'meta-search engines' (search engines that search the search engines) such as Dogpile and Copernic useful in some contexts. However, be aware that if you use simple search terms you may end up with more results than you can cope with. You will need to perform Boolean searches using AND and OR to narrow the output to a reasonable number of results.

8.8.1.1   *Case study   Virtual Training Suite*

www.vts.intute.ac.uk/: the Virtual Training Suite is a set of free, subject-specific tutorials that help students develop their internet research skills and find relevant information quickly. Each tutorial lasts for approximately one hour but can be completed in sections, as time permits. The tutorials are available for a range of different subjects, each written by lecturers or librarians with in-depth knowledge of the subject.

For the biosciences, there are several relevant tutorials:

- Agriculture
- Biodiversity
- Medicine
- Microbiology
- Pharmacy
- Plant sciences
- Veterinary medicine.

Case study by kind permission of the Virtual Training Suite.

8.8.2   *Getting information to come to you*

You can keep up to date with new information published online, be it news or scholarly papers, using RSS (Really Simple Syndication; see Section 8.12, Additional resources). This is a service that will provide you with an 'RSS feed' of updates and changes made to a website; all you have to do is subscribe to the RSS facility at that website. You can 'read' an RSS feed using an RSS reader such as Google Reader (reader.google.com). An RSS reader is a single website that can contain all of your RSS feeds; the feeds are automatically updated and new information highlighted. This means that you can instantly be aware of updates to as many websites as you wish, all at a single location.

Another simple way to access RSS feeds quickly is by using a 'widget' on a personalised homepage such as iGoogle (www.google.com/ig) or Netvibes.com (www.netvibes.com). A personalised homepage will help you make the most of the Worldwide Web, and you can create a widget that will display your latest RSS feeds each time you access the Internet. You can also create widgets that will allow you to interact directly with other websites. For example, widgets that allow you to update your Facebook postings or Twitter status. In addition, you can include a link to a task manager such as 'Remember the Milk' (www.rememberthemilk.com) or a direct link to your Google calendar to remind you, for example, when and where your next lecture will take place.

You can search for RSS feeds to subscribe to using a search engine (try 'RSS biology'), or go to the BBC news website for information on how to subscribe to their Health or Science news feeds. *New Scientist* also offers a great range of RSS news feeds which you can select by topic. Look for the orange and white RSS icon on any web page that you regularly visit; many sites offer RSS.

8.8.3 Academic journals online

Whilst 20 years ago you would have been expected to read print copies of journal articles and textbooks for your studies, the move to online publishing means that you now have instant access, from your desk, to a very large amount of published literature including journals describing the latest scientific research. The number of research papers published per year has steadily increased, and a recent report showed that the number of papers read by scientists each year has doubled during the last 30 years (Fry *et al.*, 2009). Students who demonstrate an awareness of current developments in their field of study are often richly rewarded, particularly during marking of final-year exam papers. It is therefore essential that you keep up to date with what is happening in your subject, and you can do this by adopting strategies that will keep you abreast of the latest developments. For example, you can make use of RSS feeds, made freely available by publishers, that provide information about the contents of the latest edition of a journal. This means you will have immediate access to the title, authors and short abstract for each paper contained within the journal. RSS feeds can also be customised using specific search terms to help you identify relevant published material. For example, you can use PubMed to search for a particular author or keyword then click on 'RSS' above the search box to subscribe to a customised feed based on that term (see http://tinyurl.com/pubmedh).

You may just want to keep abreast of more general information. There are several journals that provide extremely useful summaries of recent advances in many areas of science. They include:

- *Scientific American*
- *The Current Opinion* series, i.e. *Current Opinion in Biotechnology* ...–... *Current Opinion in Structural Biology*
- *The Trends* series, i.e. *Trends in Biochemical Sciences* ...–... *Trends in Plant Science*.

Subscribing to the RSS feeds for these journals will provide an easy way to keep up to date and broaden your subject knowledge.

8.8.4 Textbooks online

Tutors and lecturers will recommend textbooks for their courses. While personal paper versions of texts are useful because you can annotate (make notes, highlight) your own copy, it is always worth a quick Internet search to establish whether the texts are available free online. Besides being free, further advantages of online textbooks are that they are available anytime, anywhere, and they are fully searchable.

The number of freely available eBooks available at Google Books (books.google.com) is constantly increasing, and some seminal historical texts now out of copyright (such as Charles Darwin's essays and books) are available from open-source projects such as Project Gutenberg (www.gutenberg.org). There are free books available from iBooks accessible via Apple iPods or iPhones. Another invaluable source of online textbooks is the National Library of Medicine in the USA (www.ncbi.nlm.nih.gov/books); their bookshelves house full text versions of most of the genetics textbooks you will need along with some key biochemistry and biology texts (*Biochemistry* by Berg, Tymoczko and Stryer; *Genomes* by Brown; *Molecular Biology of the Cell* by Alberts *et al.*). These books are fully searchable and contain all of the diagrams and other images from the paper copy. They are usually only one edition earlier than the current print edition but, as most textbooks change very little between editions, this is likely to be sufficient for most needs in the early years of your course.

8.8.5 Collecting and sorting your resources

Often, we find useful articles or video clips but afterwards forget where they are located. The 'bookmarking' system available with the browser on your computer will allow you to organise and readily access these resources. However, you will only have access to these bookmarks when you are using that particular computer. When you are working at university you may use several computers in one day; for example, computers in the library, the laboratory, in halls of residence, or your own laptop or smartphone. When you go home during the vacation and want to revise or work on a dissertation, you will need to have your bookmarks with you. One way to do this is to use a portable browser, such as Chrome or Firefox (http://portableapps.com/). These can be stored on a USB stick and used with all of the computers you utilize. Another option is to store bookmarks online at bookmarking sites such as Delicious (www.delicious.com; other examples are diigo, www.Diigo.com, and Google bookmarks, www.google.com/bookmarks/). These sites enable you to add bookmarks, with notes, and organise them using a series of tags (short words or labels) so that you can build up a set of resources online. For example, you could use a tag for each of your modules, or create one when you are working on an essay. When you have completed your research for the essay you can then go to the Delicious site where you will see all the websites you marked with the tag for the essay.

An advantage of using a system such as Delicious is that you can benefit from access to other people's bookmarks by subscribing to their Delicious accounts. You could, for example, share bookmarks with your peers when you are all working on a particular group assignment. Delicious creates RSS feeds on all of its pages; so, if you want to keep track of new developments in a particular area, you can follow the pages bookmarked by the Delicious community via the RSS feed for a single tag (for example, http://delicious.com/tag/biochemistry).

8.8.6 Case study   Social bookmarking

A compulsory first-year ICT and numeracy module for biological scientists at the University of Leicester requires students to share bookmarks on the social bookmarking service Delicious (www.delicious.com). As students come across journal articles, YouTube videos or news clips relevant to their studies, they can save the URLs for these sources in their Delicious account and categorise them by adding tags. Tags are words that describe the content of the web page. For example, as a student of microbiology you might save a BBC article on a new vaccine for TB under the tags 'TB', 'disease', 'immunity', 'vaccine'. Any user of Delicious could then search for all of the bookmarks saved under these tags and would find the articles you had bookmarked along with bookmarks saved by others under these categories. This approach to social bookmarking can also be used by students on the same module. If they create a tag that is specific for the module, for example by using the module course code and year ('uolbs1009'), they can search for items related specifically to that module. To see an example, look at the module-specific tag uolbs1009 (http://delicious.com/tag/uolbs1009-09). Here, individuals have contributed to a wide range of bookmarks for each of their first-year modules. The students may find these resources particularly useful at revision time.

8.9   Working effectively

8.9.1   Typing efficiently

Your university will most likely require you to submit coursework as word-processed documents. Working at a keyboard is an essential skill and, if you are prepared to take some time to learn how to touch type, this will help you to work more effectively. Online tutorials are available that will help you improve your typing skills (http://keybr.com/, www.bbc.co.uk/schools/typing/). Alternatively, ask staff at your university's Student Learning Centre or Library if they offer touch-typing courses.

8.9.2   Managing your references

A very important part of your development as a life scientist involves learning from your colleagues and predecessors. By taking a life science degree, you become part of the academic scientific community. Good scientists challenge and critically evaluate the work of others. Since the academic community builds on previous work to create new knowledge and understanding, a system of trust and acknowledgement has been developed to allow scientists and academics to relate their work to the work of others. During your studies you will use textbooks and read papers in which scientists discuss thoughts that they have had and, perhaps, theories they propose. When you are writing about thoughts and proposals that originate from others, you need to be careful to acknowledge the source of these ideas. Presenting such work as your own, without acknowledgement, is plagiarism (see Chapter 9). One day, you may add to the body of knowledge by contributing new theories that are explored through experiments, and you will want to be given credit for your hard work. As a student it is therefore very important to make sure that you acknowledge the work of other scientists by using the referencing and citation system effectively.

Referencing is the way to record the sources of information that you have used when writing a piece of work. You compile a list of sources at the end of the work and refer to them in the main text using *in-text citation*. For example:

'The structure of DNA was deduced by two Cambridge scientists (Watson & Crick, 1953).'

Reference:

WATSON, J.D. & CRICK, F.H.C. (1953). Molecular structure of nucleic acids: a structure for deoxyribose nucleic acid. *Nature* **171** (4356), 737–738.

There are many different sources of journals and other publications you may want to use for your work, and each publication follows its own particular format for in-text citations and references listed at the end of an article. This can be rather confusing, but your tutors will advise you on the accepted style for the coursework they set. However, you will need to recognise the wide variety of systems in place. One way to keep the salient information for each reference readily at hand, and to be able to re-use and cite it in a variety of different ways, is to make use of a reference management system. Reference management systems will help you to record the right information about each reference: the author(s), date of publication, full journal title and so on. They will then allow you to format the references in the style of your choice. Some universities provide styles and templates to use in reference management systems and these ensure that your references are always in the correct format for coursework. You will find these systems particularly useful when you are preparing a dissertation that involves a review of a large number of publications.

There are two main types of reference management system. The first involves software that is loaded onto your desktop computer and that integrates with Word or other word-processing systems. You insert relevant, in-text citations as you write, and the software automatically compiles a reference list of these sources at the end of the work. Alternatively there are online systems, which may or may not allow in-text citation, but which have the advantage of access anywhere, anytime and beyond the end of your degree studies. A further advantage of online systems is that they can incorporate a social element.

8.9.2.1 Desktop-based reference management software

A widely used proprietary software package for reference management is Endnote (www.endnote.com), and your university may provide access to this software. Endnote has full integration with Microsoft Word. If you use a university-licensed copy of Endnote you can export your references at the end of your course and use them in one of the free online reference managers.

8.9.2.2 Online reference management systems

Refworks (www.refworks.com) is a subscription-based service that may be provided by your university. This online system works in a very similar way to Endnote and, depending on your university's licensing arrangement, may enable integration with Microsoft Word. One advantage of using Refworks and other online referencing systems is that they provide *bookmarklets* that can be added to your browser so that, when you find a paper or other publication of interest, you can very quickly add the information to your library. The bookmarklet will convert all the citation information into a new record, removing the potential for errors or misinterpretation that can occur when recording a reference manually.

Citeulike (www.citeulike.org), Connotea (www.connotea.org) and Mendeley (www.mendeley.com) are examples of free-to-use online referencing systems. They all provide tools similar to those provided by Refworks, including bookmarklets that enable import of references into your library in a wide variety of styles. These three systems have the added advantage that they allow you to build relationships, online, with other users. Like other social bookmarking sites they offer tagging and sorting facilities. By monitoring papers bookmarked by all users in real time they can also provide information about which papers are most popular and presumably perceived as most important by the scientific community. This form of social citation software is in its infancy but is rapidly growing in popularity. Citeulike will also recommend other papers that might interest you based on the library of references you have accumulated (just like the Amazon system that recommends books to read based on your past purchases and viewing habits!).

8.9.2.3   Case study   Don't cheat yourself!

http://www2.le.ac.uk/offices/ssds/sd/ld/resources/study/plagiarism-tutorial

University guidance on plagiarism is often expressed in legal terms and may appear threatening, as the penalties for plagiarism can be severe. Stuart Johnson of the Student Support and Development Service at the University of Leicester therefore devised this online tutorial (Figure 8.3) as a source of friendly, helpful and practical advice. The tutorial is designed to help students recognise that avoiding plagiarism will improve the standard of their academic work. It includes the views and conduct of four 'students' who plagiarise for one of the following reasons:

- poor time management;
- difficulty in finding their own voice and thoughts on a subject;
- poor note-taking skills;
- lack of understanding of the referencing and citation system.

The tutorial uses the four case studies to demonstrate to students how plagiarism can easily occur. It includes an interactive exercise where students can test their knowledge of whether or not sample essay extracts in their subject area contain plagiarised material. The tutorial makes practical suggestions on how to: take notes effectively; plan/set aside time for researching and writing essays; get advice on using referencing systems.

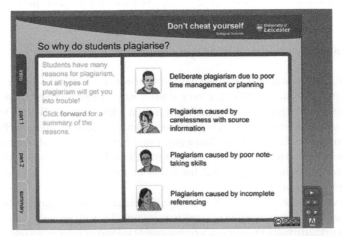

Figure 8.3   Screenshot from *Don't Cheat Yourself* tutorial showing interactive exercise. Taken from www.flickr.com/photos/39702960@N07/4896679263/. With kind permission from Stuart Johnson, Student Support and Development Service, University of Leicester

Your university will probably have its own online guide to plagiarism that describes what it is, how to avoid it and the penalties that might be incurred. See, for example, http://skills.library.leeds.ac.uk/avoiding_plagiarism.php.

8.9.2.4 Tutor notes

It is important not only that penalties for plagiarism are made explicit, but also that clear guidance is given to students regarding how plagiarism can be avoided, including help with citation and referencing practice. Tutors should recognise that many students arriving at university have not developed these skills and may not appreciate the academic necessity of attributing sources of information and distinguishing them from material they have created themselves. Practical exercises accompanied by formative feedback are invaluable in promoting good academic scholarship and the avoidance of plagiarism. It is important that students are given a clear and consistent message regarding academic integrity and the positive benefit of this for their own development.

The now widespread use of electronic detection systems, such as Turnitin, CopyCatch and SafeAssign (see Section 8.12, Additional resources) does not negate the need for academic judgement when assessing cases of potential plagiarism. In addition, it should be borne in mind that students may have permission to check their own work through detection systems in advance, or may have used commercial services e.g. Writecheck.com; so consistent advice must be made available at all times. For advice and guidance on designing coursework that allows little opportunity for students to plagiarise, see Additional resources.

8.10 How you can achieve your potential using computers and online resources

- Ensure you are fully aware of all of the search facilities, databases and other resources available through your university's portal and VLE.

- Use electronic/online diaries, calendars, 'to do' lists etc. to organise your time efficiently.

- Make use of the technologies, outlined in this chapter, that ensure information is channelled effortlessly *to you*.

- Build a Personal Learning Environment.

- Learn how to establish a professional online presence/reputation/identity.

- Use discussion boards, social networks and other Web 2.0 technologies to support your studies, but establish separate private and professional identities.

- Make the most of freely available open-source materials.

- Make full use of reference management systems.

- Utilise appropriate procedures, e.g. password protection, that ensure your electronic resources are safe and secure.

- Ensure you know what plagiarism is and how you can avoid it.

- Be aware of legal issues, e.g. copyright, that may apply to materials you access online.

8.11 References

e-skills UK (2008) Technology Counts: IT & Telecoms Insights 2008. e-skills UK Sector Skills Council (www.e-skills.com/Research/Research-publications/Insights-Reports-and-videos/Insights-2008/9).   Accessed April 2011.

Fry, J., Oppenheim, C., Probets, S. *et al.* (2009) PEER behavioural research: authors and users vis-à-vis journals and repositories. *Baseline Report,* Loughborough University (www.peerproject.eu/fileadmin/media/reports/Final_revision_behavioural_baseline_report_20_01_10.pdf). Accessed April 2011.

Hoadley, C.M., Xu, H., Lee, J.J. and Rosson, M.B. (2010) Privacy as information access and illusory control: the case of the Facebook News Feed privacy outcry. *Electronic Commerce Research and Applications* **9**, 50–60.

8.12 Additional resources

8.12.1 Books

Johnson, S. and Scott, J. (2009) *Study & Communication Skills for the Biosciences*. Oxford: Oxford University Press.

Gash, S. (1999) *Effective Literature Searching for Research*. Farnham: Gower Publishing.

Shields, G. and Pears, P. (2008) *Cite Them Right: The Essential Guide to Referencing and Plagiarism* (7th Edition). Newcastle upon Tyne: Pear Tree Books.

8.12.2 Online resources (accessed April 2011)

Skills in Accessing, Finding and Reviewing Information, Safari. Open University. www.open.ac.uk/safari/.

The Internet Detective. Intute. www.vts.intute.ac.uk/detective/.

E-skills UK – website of the Sector Skills Council for Business and Information Technology. www.e-skills.com/.

E-learning in the Disciplines. JISC/Higher Education Academy report. www.heacademy.ac.uk/assets/York/documents/ourwork/learningandtech/elearninginthedisciplines.pdf.

The Higher Education Academy Centre for Bioscience website. www.bioscience.heacademy.ac.uk/.

Case studies of e-learning: www.bioscience.heacademy.ac.uk/resources/themes/elearn/elearncs.aspx.

ImageBank: www.bioscience.heacademy.ac.uk/imagebank/.

JISC Techdis – provides a wide range of resources, advice and information on the use of assistive technology in teaching for Higher Education. www.jisctechdis.ac.uk/.

Information page for staff involved in the development of e-learning content: www.jisctechdis.ac.uk/techdis/technologymatters/contentcreation/creatinginclusivecontent.

Information on how to create more accessible learning resources using Microsoft Word, PowerPoint and PDFs: www.jisctechdis.ac.uk/accessibilityessentials.

RSS – there are helpful introductions on the BBC website (www.bbc.co.uk/news/10628494) and at www.squidoo.com/rss-explained. Alternatively you may want to watch the short video at www.commoncraft.com/rss_plain_english.

8.12.3 Software (accessed April 2011)

Thunderbird. Open-source email client, free to download, enables you to read and reply to emails on your desktop. www.mozillamessaging.com/en-US/thunderbird/.

Software packages for carrying out statistical analysis including regression, ANOVA and experimental design:

Minitab www.minitab.com/.
Stata www.stata.com/.
R (free, open-source software) www.r-project.org/.
SPSS www.spss.com/uk/.
(All except R will need to be purchased, but check if your University offers any of these under licence on campus before buying.)

8.12.4   Web browsers (alternatives to Internet Explorer, all free)

Opera. Works well on mobile devices. www.opera.com/.
Mozilla Firefox. Add extensions easily to work more efficiently online. www.mozilla.com/en-US/firefox/.
Chrome. Browser from Google, very clean interface. www.google.com/chrome.

8.12.5   Plagiarism detection and avoidance (accessed April 2011)

Turnitin. Compares work submitted by students to an extensive database containing: work previously submitted by students, website content, peer-reviewed journal papers and eBooks. Institutional subscription required. www.submit.ac.uk/.
CopyCatch. Compares work submitted by students and identifies collusion; can also indicate if material is likely to have been copied from other sources. Institutional subscription required. www.cflsoftware.com/.
SafeAssign, now hosted within the VLE Blackboard. SafeAssign is similar to Turnitin: it compares student work with sources on the Internet. http://safeassign.com/.
Student guides providing advice on how to avoid plagiarism:
http://www2.le.ac.uk/offices/ssds/sd/ld/resources/study/plagiarism-tutorial.
http://skills.library.leeds.ac.uk/avoiding_plagiarism.php.
Tutor guides, and case studies, on designing out plagiarism: www.plagiarismadvice.org/resources/designing-out-plagiarism.

9 Bioethics

Chris J. R. Willmott

9.1 Introduction

Imagine that you've signed up to do a bioscience-related degree course (I hope, given the fact that you are reading this book, that this is not too difficult a stretch of the imagination). As you sit in the lecture theatre chatting with your neighbour, a man, whose terrible dress sense betrays immediately that he must be an academic, comes into the room and starts talking with gusto about his research into terracotta figurines of the late Bronze Age or perhaps about the importance of authorial subjectivity in sixteenth century poetry. With some justification you might conclude that you've misread your timetable and gone to the wrong lecture theatre.

Now imagine instead that the lecturer had launched into consideration of the rival philosophical viewpoints espoused by Immanuel Kant and John Stuart Mill; would you still feel that you were in the wrong place? My aim within the next few pages is to try to demonstrate that, far from being an irrelevant distraction, having an understanding of ethics and moral philosophy will actually enrich your studies as a bioscientist (although, as we'll see below, I don't believe that contrasting Kant and Mill is the best place to start!).

9.2 The rise of ethics in the bioscience curriculum

My conviction that discussion of ethics has an important role to play within science education is shared by a growing number of people. Recent curriculum developments around the world, but particularly within the UK, have emphasised the need to reflect upon the ethical and moral impact of scientific innovations (Willmott and Willis, 2008). For example, the National Curriculum for England and Wales includes as a foundational concept the importance of 'Examining the ethical and moral implications of using and applying science' for all pupils over the age of 11 (QCA, 2007).

At university level, the subject benchmarking statement for bioscience published by the Quality Assurance Agency in 2002, and lightly revised in 2007, outlines the expected content of all bioscience courses in the UK (QAA, 2007). The statement contains recommendations regarding

Effective Learning in the Life Sciences: How Students Can Achieve Their Full Potential, First Edition.
Edited by David J. Adams.
© 2011 John Wiley & Sons, Ltd. Published 2011 by John Wiley & Sons, Ltd.

the opportunities for undergraduate bioscientists, in whatever field of study, to develop their ethical thinking. These include:

- an expectation that students will 'be confronted by some of the scientific, moral and ethical questions raised by their study discipline, to consider viewpoints other than their own, and to engage in critical assessment and intellectual argument' (Section 3.1);

- bioscience students, the statement continues, should be able to: 'recognise the moral and ethical issues of investigations and appreciate the need for ethical standards and professional codes of conduct' (Section 3.5); and

- 'undertake field and/or laboratory investigations of living systems in a responsible, safe and ethical manner' (Section 3.6);

- by the end of their degree, a typical honours graduate emerging from any bioscience programme should 'be able to construct reasoned arguments to support their position on the ethical and social impact of advances in the biosciences' (Section 5.8).

One reason, therefore, that university courses in biology and related disciplines are likely to include discussion of ethics is because programme directors have been told to ensure that this is the case! The motivation for developing ethics content in a bioscience course is not, however, simply an issue of rule-keeping. In this regard, it is useful to consider why emphasis on ethics was included in the benchmarking statement in the first place, and how science students who are mindful of these issues may actually become better scientists as a result.

9.2.1   Read all about it!

As moral philosopher Stanley Grenz has observed 'Modern Science has placed in our hands capabilities that have aggravated long-standing ethical problems as well as introducing new quandaries' (Grenz, 1998). To recognise the importance of bioethics we actually need to look no further than the headlines of our newspapers and television news bulletins. In any given week there are likely to be several stories which starkly illustrate that science does not proceed in a value-free vacuum. Knowing I was due to write this chapter, I kept a careful eye on stories covered during a two-week period (an activity I have conducted periodically for a number of years and observed on each occasion a similar breadth of issues).

In the two weeks in question, the news included: discussion of pre-implantation genetic diagnosis (PGD) and 'designer babies'; concern about an ecological crisis; provision of fertility treatment to post-menopausal women; worries about use of nanotechnology in the food industry; genetically modified crops; and fraud in scientific publications. Whatever particular flavour of bioscience you are interested in, this brief list likely contains some topic related to your studies (and if not, perhaps one of the other 20 stories I omitted from the list for clarity would have done so). This breadth of coverage is the norm and not in any way exceptional.

9.2.2   I'm sorry I haven't a clue

Hand in hand with an explosion of ethically contentious developments in bioscience, there have also been concerns that the public in general are not well equipped to understand, and therefore evaluate the validity of, the stories under consideration. What, for example, is a 'designer baby'? What are the technological issues associated with selection of a particular embryo? And what are the ethical consequences of any decisions made?

In recent years there has been lively debate about the need for scientists to engage with the public about the significance of their research and innovations. No-one questions the underlying relevance of the process, but there has been some contention regarding whether scientists ought to be *telling* the public about the science (labelled the 'deficit' model) or *discussing* it with them (the 'dialogue' model) (Wilsdon and Willis, 2004). Whichever is preferable, one example will show why it is especially important that you as a student of bioscience have a good grasp of both the science and ethics of current developments.

Ever since 1973, the European Commission has conducted periodic surveys of public opinion within the various member states. In one of these so-called *Eurobarometer* surveys, they asked members of the public if they agree with the statement 'Ordinary tomatoes do not contain genes while genetically modified tomatoes do.' In response, 35% of Europeans said that they agreed with the statement, and a further 30% said they did not know. In other words, nearly two thirds of the people polled were not confident that ordinary tomatoes possessed genetic material.

Faced with this level of ignorance, it is important that bioscience graduates, at the very least, have a good grasp of the issues involved. Added to this, many readers of this book may actually go on to take an active role in scientific research and therefore it is all the more important that they (you?) have given careful consideration to the appropriateness of a line of scientific inquiry.

9.3   What exactly is bioethics?

Although we have started to consider some examples of specific biomedical and ecological developments where ethics plays a prominent role, we have thus far avoided a definition of bioethics as a discipline. It is possible that some readers, particularly those from America where the term 'bioethics' is frequently synonymous with 'medical ethics', will be wondering about the breadth of issues that might be included.

In the context of this chapter we will use the term 'bioethics' to encompass three distinct (though overlapping) areas: these are biomedical ethics, environmental ethics and research ethics. This broader definition is in keeping with the spirit of its usage by American cell biologist Van Rensselaer Potter who, it is generally acknowledged, was the first to use the word 'bioethics' to describe 'a new discipline which combines biological knowledge with a knowledge of human value systems' (Potter, 1971). Let's take a closer look at the three dimensions of bioethics.

9.3.1   Biomedical ethics

Many aspects of healthcare have clear ethical dimensions; though not all of these are equally relevant within bioscience courses. One commentator (Cameron, 2004) has advocated a classification of biomedical ethics issues into three subtypes. Firstly, there is *Bioethics 1*; these are issues such as abortion and euthanasia which are still highly relevant, but have been possible, and hence discussed, for many centuries. Secondly, there is *Bioethics 2*, which covers issues such as PGD and the use of stem cells. These are relatively new developments that are either already being used clinically, or are deemed likely to be used in the near future. Thirdly, there is *Bioethics 3*, which includes potential development in fields such as artificial intelligence and the creation of nanoscale robots that may, for example, be able to circulate within the bloodstream aiding the battle against infection. This kind of development is more speculative and may not have medical relevance for some time to come. One attraction of this three-part division into Bioethics 1, 2 and 3 is the fact that they equate respectively to 'taking life', 'making life' and 'faking life' as a helpful aide memoire.

9.3.2 Environmental ethics

If biomedical ethics has most direct relevance to students with an interest in molecular biology, environmental ethics may have greater bearing for students within more whole-organism disciplines. Concern about the environment can be expressed with a variety of ethical motivations. At one end of the spectrum, which one author has labelled *light green ethics* (Curry, 2006), interest in ecology arises from its potential impact on mankind; in other words it retains an anthropocentric focus. From this perspective, global warming, sustainability, and concern about loss of habitat are of ethical relevance because they may end up affecting humanity. A recent headline on the BBC news website, 'Nature loss "to damage economies"', illustrates this type of thinking (Black, 2010).

At the other end of the scale, *dark green ethics* would be characterised by a more holistic or ecocentric approach in which the integrity of ecosystems and the planet as a whole are of maximum value. For those adopting this model, a clash between human self-interest and wider ecological interests retains the possibility, or even the probability, that correct resolution of the conflict will go against **human** interests.

Between these two extreme positions would be a variety of other frameworks, such as biocentrism, a philosophical view in which respect should be afforded to all species, of which humans are but one and in no way superior to the others (except, I suppose, in their capacity to do harm).

Specific issues falling under the umbrella of environmental ethics might include the cultivation of genetically modified crops, management of woodland ecosystems, decisions to switch from growth of food crops to production of biofuels, or the introduction of non-native species to control a particular biological menace (frequently, as in the case of Japanese knotweed, another non-native species introduced in an earlier era!).

9.3.3 Research ethics

The case of disgraced stem cell scientist Hwang Woo-Suk is probably the most notorious recent example where significant failing in the ethical conduct of research has been brought into the glare of public attention. Hwang went from hero to zero in his native South Korea when it transpired that he had fabricated many of his ground-breaking results in therapeutic cloning (the plan to 'personalise' cells for treating a patient by removing the nucleus from an embryonic cell and replacing it with the nucleus from one of the recipient's own cells).

Such obvious, and profound, deviations from acceptable practice may be rare, but there are a plethora of other issues which might be considered 'research ethics'. As in other walks of life, ethical norms associated with research have generally been assumed rather than spelt-out specifically, and have tended to come into sharp focus only at times when they have been clearly violated.

Not telling the whole story

Along with the outright faking of experiments, there is the possibility that a scientist might make a conscious decision to selectively report only the results that fit with their predetermined hypothesis and hide data that tell a different story. During the so-called 'Climategate' saga prompted by the leaking of previously private e-mails in November 2009, it was claimed that there had been a deliberate attempt to suppress publication of data that contradicted the majority view that human activity has influenced the temperature of the planet (Hickman and Randerson, 2009).

The conduct of scientific research is actually rather more complex than might initially meet the eye. Notice in the previous paragraph that I purposefully emphasised that the decision was *conscious* and *deliberate*. It is part of human nature to assume that results that fit our hypothesis are more likely to be 'correct' than those that go against our expectation. We may therefore subtly, and *subconsciously*, place greater emphasis on data that match the hypothesis, and downplay results that do not. This kind of *self*-deception whilst not ideal is, I would suggest, a far cry from *knowing* that you have conducted an experiment 10 times but only reporting the 3 cases where the data fitted your model, and completely suppressing the existence of the other results. It is for this reason that strict rules on record-keeping and the storage of laboratory notebooks are increasingly being brought into force (Shamoo and Resnik, 2009).

Telling someone else's story

As a contemporary student, it is likely that you will have received dire warnings about the consequences of plagiarism. There are various definitions of plagiarism, but in essence they all boil down to taking the ideas or writings of somebody else – perhaps a textbook author, another student or someone paid to write an essay on your behalf – and passing it off as though it was your own original work. This over-dependence on someone else's work may arise deliberately or may occur by mistake (for example as the result of poor note-taking habits), but either way it is seriously frowned upon.

From time to time, academics also get caught out peddling other people's wares as their own. One recent UK case involved celebrity psychiatrist Raj Persaud, who was suspended from practice for three months after admitting he had plagiarised sections in his 2003 book *From the Edge of the Couch* and other articles (Dyer, 2008). If you keep an eye out when reading scientific journals, you will sometimes spot retraction of articles that turned out not to be quite as original as the corresponding authors had implied. One apology that caught my eye was in the journal *Trends in Biochemical Science* (TIBS), which noted that an article they had published a few months earlier 'contains large tracts that have been copied from a previously published *Nature Cell Biology* News and Views article' (Wilson, 2002). My curiosity was sufficiently aroused to dig out both the offending article and the source document. The editors of TIBS weren't kidding when they said 'large tracts' had been copied – a side-by-side comparison showed that most of the more recent paper had simply been cut and pasted from the original.

Although wholesale and blatant copying of this kind is unusual, the example given is not, unfortunately, an isolated case. In the same way that many universities are now using computer programmes to scan student work for plagiarism, scientific journals have been trialling a service, 'CrossCheck', for checking whether submitted articles contain material previously published elsewhere. During the trial it is reported that one journal found that 23% of submissions, which they were otherwise prepared to accept, turned out to contain plagiarised text (Butler, 2010).

Telling the same story again

Several cases identified by the new CrossCheck service were not guilty of copying other people's work, but rather of *self*-plagiarism; that is seeking to repackage material that the author(s) have already published elsewhere. Universities, government and other research funders carefully scrutinise the number of publications that scientists produce in order to justify giving them money to carry out their research and/or valuable lab space in which to do their experiments. In this context, self-plagiarism is an attempt to boost the number of articles the authors have produced since the last time someone checked up on them. If you are using Web of Knowledge or PubMed (or similar) to try and source articles for an essay or dissertation, you will sometimes uncover evidence of this sort of thing; for example, two (or more) articles published by the same three authors (possibly with the order of their names swapped around) with very similar (but slightly different) titles coming out just a few months apart in different journals. Duplicate publication of this kind is not big and it's not clever, but if someone just wants to know how many papers you've produced then it can look like you've generated more data than you actually have.

Telling a different story

Scientific research is an expensive business. Money to conduct research is usually generated via application to various government bodies or charitable groups in which you explain what you intend to do with their money. The figures involved can be substantial; many university scientists, for example, will have a total budget in excess of a million pounds with which to pay for their research.

In most cases, when a funding body has weighed your application and decided to give you money, there is a not unreasonable expectation that you are going to carry out the experiments that you promised to conduct. It would be fraudulent for you to take their cash and use it either to pay for expensive holidays and luxury items or, less dramatically, to bank-roll an entirely different research programme. Interestingly when Hwang was eventually given a prison sentence (suspended) it was for embezzlement of South Korean Government money, since he was deemed not to have done the research for which he received funding (Bae, 2009).

Conduct unbecoming?

Even before the Hwang Woo-Suk scandal grabbed headlines around the globe, there was sufficient concern about research misconduct for governments, professional bodies and scientific societies to put in place a variety of rules and safeguards. In the United States, allegations of scientific fraud at a variety of research institutes during the 1980s led to the establishment in 1989 of the Office of Science Integrity and the Office of Scientific Integrity Review, later amalgamated into the current Office of Research Integrity (ORI). Individual learned societies, such as the American Society for Microbiology (1988) and the American Society for Biochemistry and Molecular Biology (1998), adopted codes of ethical conduct to which they expected their members to adhere.

Within the United Kingdom, progress towards codes of conduct has been a little slower. In 2003, the House of Commons Select Committee on Science and Technology urged 'scientific learned societies to consider introducing an overt ethical code of conduct as a prerequisite of membership and back this up with programmes to heighten awareness of the issues involved' (Gibson *et al.*, 2003). In so doing, they were adding their weight to the call already made by Nobel laureate Sir John Sulston for a universal code analogous to the 'Hippocratic Oath' (or later equivalents) affirmed by medical doctors (Briggs, 2001).

Following a period of consultation, the clamour for a universal code of ethics for science led to the publication in 2007 of the UK Government document *Rigour, Respect, Responsibility* (DIUS, 2007), which emphasised a seven-point code for the conduct and communication of scientific research (see Box 9.1). In truth, the striking thing about each of these items is how 'obvious' and 'sensible' they are, calling into question whether they add any meaningful depth or guidance to the normal working of scientists.

The opposite perspective, of course, would be to see the need to emphasise such straightforward principles as an indication of the seriousness of the malaise within the practice of science.

Adding stick to the carrot

The major criticism of the government's Code of Conduct has actually been its lack of teeth; there is no obvious mechanism by which breach of any of these key principles would lead to penalty or consequence. In July 2009, Research Councils UK, the umbrella organisation representing the seven government-funded research councils in the United Kingdom, published *Integrity, Clarity and Good Management* (RCUK, 2009), setting out more fully the necessary standards of research

> **Box 9.1 A Universal Code of Conduct for Scientists**
>
> - Act with skill and care in all scientific work. Maintain up-to-date skills and assist their development in others.
>
> - Take steps to prevent corrupt practices and professional misconduct. Declare conflicts of interest.
>
> - Be alert to the ways in which research derives from and affects the work of other people, and respect the rights and reputations of others.
>
> - Ensure that your work is lawful and justified.
>
> - Minimise and justify any adverse effect your work may have on people, animals and the natural environment.
>
> - Seek to discuss the issues that science raises for society. Listen to the aspirations and concerns of others.
>
> - Do not knowingly mislead, or allow others to be misled, about scientific matters. Present and review scientific evidence, theory or interpretation honestly and accurately.
>
> Taken from *Rigour, Respect, Responsibility* (DIUS, 2007)

conduct, governance and training expected of institutions and individuals. Although (at time of writing) the details of sanction procedures have yet to be confirmed, the direct link between RCUK and funding should make it harder for scientists to ignore their obligations to ensure that research is conducted in accordance with appropriate ethical standards.

9.4 Putting the case for ethics education

Assuming that your science course includes a component of bioethics, you may well find that your lecturers elect to approach the main issues via a series of case studies. By way of illustration, consider the following scenario.

9.4.1 Case study

Craig and Jennifer Wyatt have been married for 11 years. They would love to have children. Unfortunately, Jennifer had breast cancer when she was 24, and although the chemotherapy has brought total remission from the disease it also caused damage to her ovaries which has left her infertile. The Wyatts have been on the waiting list at their local IVF clinic for several months awaiting donated eggs to try and have a baby. At present, however, there are 200 potential mothers seeking each donated egg, and the couple know that realistically they may never receive a donated egg via the normal channels.

Researchers at the hospital attached to the IVF clinic have recently been granted permission to carry out experimental procedures using eggs harvested from aborted fetuses. The technique relies on the fact that female fetuses already contain all of the undeveloped eggs that would ultimately mature during their adult lives. The researchers are looking into ways to artificially ripen these eggs as a way to overcome the shortfall in other available eggs. Her gynaecologist suggests to Jennifer that she may want to volunteer as one of the recipients of the newly developed eggs. Unsurprisingly, this approach is controversial, but for Jennifer it may represent her only chance to receive a donated egg.

> *You are a friend of Jennifer and Craig. Take a few moments to think about the range of issues raised by this case. Try to consider the situation from a variety of perspectives, including reasons for and against their involvement in this research.*

I've used this case as part of an introductory lecture on bioethics for a number of years; it's fictional, but only marginally so (the statistic about 200 people wanting each donated egg is real, and researchers are genuinely considering the potential to utilise eggs from fetuses (Hutchinson, 2003), but this work has not yet proceeded into any clinical trials).

So what issues did you come up with? Although students frequently impress me by bringing fresh insight, there tends to be a consistent core of observations made each time this case is discussed. A lot of people raise the issue of rights: does the woman who had the abortion, and is biologically the 'grandmother' of any new child, have any say in whether this research is permitted or any right to future contact with the child? Does the fetus itself have any rights? They also question the right of scientists to manipulate nature in this way. For some this development just sounds disgusting – the very notion of it elicits a 'yuk' response. These views all have something in common, which we'll discuss in a moment, but for now let's call them Type A.

Other people find themselves wondering about the likelihood that the process will be a success. What happens if the experimental treatment goes wrong and, for example, the new baby survives but is severely handicapped? Although this is a different technology, they may argue, for example, that there were 277 failed attempts before Dolly the Sheep was successfully cloned (Arthur, 2003), and there may be similar difficulties whilst perfecting this new approach. Can Craig and Jennifer sue the clinic, or disown the baby, if they are presented with a malformed infant (even if it was produced using Craig's sperm)?

Suppose the process is successful and a child is produced – will it be told about the unusual circumstances of its conception? What will be the psychological impact of discovering at a later stage that his or her biological mother never existed as an independent being, that she had (for whatever reason) been rejected by *her* own mother before birth?

What if the technique is a runaway success and becomes common? Since the whole purpose of this approach was to overcome a shortfall in the availability of eggs, it is reasonable to presume that several children will be produced from the eggs within any one fetus. What, therefore, are the potential genetic consequences of the fetuses' DNA becoming over-represented in the gene pool? What safeguards would or could there be to stop two people who are biologically derived from the same fetus marrying each other and producing, unbeknownst to them, children who may share a significant number of recessive alleles and have an increased risk of manifesting inherited conditions? If the latter scenario seems far-fetched, there is unfortunate precedent in the real example of a brother and sister, twins conceived naturally but both put both up for adoption, who were reported in 2008 to have had their marriage annulled after inadvertently ending up together (Yeoman, 2008). Again each of these arguments shares something in common, so let's call them Type B.

The case involving Craig and Jennifer may seem peculiar, but it actually serves as a very useful launch pad for considering the philosophical aspects of bioethics and ethical decision-making. If we take a step back and think in general terms about how we can decide if a certain course of action is right or wrong, then there are essentially two places where we can start our deliberations: either we can put emphasis on the underlying principles and duties, the *a priori* framework that we bring to the question, or we can focus on the potential consequences and outcomes of the action. Looking back at previous paragraphs, you may have noticed that the 'Type A' arguments were grounded on first principles, rights and duties, whereas the 'Type B' arguments placed greater significance on the results and effects of the procedure.

9.4.2 Do you want moral philosophy with your bioethics?

You can actually add quite a lot to your understanding of the ethical significance of developments in biology and biomedicine simply by knowing that there are these two general criteria, i.e. rights/duties and consequences/outcomes, for evaluating whether a new procedure ought to be permitted. A good student of bioscience, however, ought to be willing to take their consideration of these ethical positions to a rather deeper level.

It would be possible to fill an entire book of this length with discussion of bioethics alone; indeed several excellent and accessible books on the subject already exist (works by Bryant *et al.* (2005) and Mepham (2008) are especially recommended). Within the constraints of the present, more broad-ranging volume, however, it is valuable to at least offer an introduction to some of the major strands of ethical thinking which started to emerge in the case study.

We have seen thus far, that ethical decision-making can place emphasis on duties/first principles or on the anticipated outcomes of an action. You will also recall that we started this chapter by imagining a lecture in which the tutor had launched into reflections on the philosophies of Immanuel Kant and John Stuart Mill. If, having thought about Craig and Jennifer's dilemma, I add that the distinction between what I termed 'Type A' and 'Type B' responses effectively describes the major differences between the philosophical thinking of these two great minds, then I hope you may begin to see why I suggested that some reflection on moral philosophy may have more direct relevance to bioscience than was initially apparent.

Kant, for example, argued that our decisions ought to be governed by an over-riding responsibility which he termed the Categorical Imperative, and which can be phrased as 'do your duty'. Arguments starting from this position are termed *deontological*. On the other hand, Mill, one of the founding fathers of utilitarianism, argued that actions were right if their outcome led to a net increase in happiness. Arguments based on outcomes are said to be *teleological* or *consequentialist*, and utilitarianism is probably the best known consequentialist philosophy (although others do exist).

9.4.2.1 Deontology

There are strengths and weaknesses with both pure deontological and consequentialist positions. The emphasis on fundamental rules can make deontological arguments seem very black and white and therefore easy to apply. The reality of life, of course, tends to be a lot messier. Where, for example, do we obtain our rules? Religion is sometimes thought of as the source of rules, but it is self-evident that adherents to different religions have different emphases, and many people would argue that they personally are not followers of any religion. Do we get rules then from something more innate and universal, conscience say or something within our genetic heritage? Even though we have a shared humanity, our 'moral compasses' are not uniformly aligned. Kant, as we have already noted, tried to get around this difficulty by reference to his Categorical Imperative to do your duty, but this just shifts the question to 'How do we know what our duty is?'

There are other difficulties with a deontological approach. How, for example, do you decide what to do if two or more general rules appear to be in conflict? Let's imagine it is 1943 and you are pottering about in your Amsterdam home when there's a knock at the door. You go to answer it, and find in the street outside members of the German security police. After a curt greeting, their commanding officer asks you if there are any Jews hiding in the house. What are you to do? You have always been brought up to tell the truth, yet you know full well that there are two families living in your loft. Do you say 'Yes there are; if you push the wardrobe in the front bedroom to one side you'll find a secret staircase – you can't miss it'? This sounds like an absurd response; you would almost certainly weigh preservation of life above truth-telling in this case. In doing so, however, we are essentially saying that the rule to tell the truth can sometimes be trumped; in which case, can it be considered an ethical norm at all?

9.4.2.2 Consequentialism

Aware of these kinds of issues when trying to start with rules and duties, Jeremy Bentham, John Stuart Mill and others argued that **outcomes** were more appropriate criteria to judge the appropriateness of an action. Our conduct would be right if it led to the greatest balance of good over evil. This initially sounds appealing, and utilitarian models in particular are employed broadly in, for example, healthcare policy-making.

As with deontological approaches, however, strict application of consequentialist philosophy soon meets problems. In the first instance, does the end justify the means? 'Suppose we could slightly increase the collective happiness of ten men by taking away all happiness from one of them,' asked philosopher A. C. Ewing, 'would it be right to do so?' (Ewing, 1953). According to some strict utilitarian models this would be fine. This scenario may not be quite as abstract and theoretical as we'd like to think, with accusations made that political prisoners in China have frequently been killed in order for their organs to be donated to more 'worthy', or more wealthy, recipients (Thomasma, 1997).

Secondly, consequentialist models suffer from the inevitable difficulty that the real outcome of an action can never be guaranteed in advance; we may have a *good idea* about what will happen, but the *reality* may prove to be quite different. The litany of disasters in which attempts to control biological pests, via the introduction of non-native predator species, have gone devastatingly awry should be warning enough that outcomes can deviate from expectation (e.g. Simberloff and Stiling, 1996). One of the best known examples involves the introduction to Queensland, Australia of the cane toad. Imported from Hawaii during the 1930s in order to suppress the population of beetles living amongst the sugar cane, the toads not only manifest a preference for eating birds' eggs, rival frogs and insects *other than* the intended beetles, but also proved fatally poisonous to various native predators, including crocodiles (Coady, 2009).

9.4.2.3 Virtue ethics

Partially as a result of the difficulties apparent with both deontological and consequentialist approaches, there has been a recent resurgence of interest in a different approach. Virtue ethics places greater emphasis on individual character and, as such, draws its heritage from both classical Greek thinkers, especially Aristotle, and from a Judaeo-Christian worldview. In some sense, virtue ethics can be seen as an *ethics of being* rather than an *ethics of doing* (Eriksson *et al.*, 2007). When push comes to shove, however, even a virtue ethicist needs to decide how they are going to act under given circumstances, and therefore some tool or tools for formulating a response remain important.

9.4.2.4   Beauchamp and Childress' Four Principles

One widely used decision-making strategy is termed *principlism*, or the Four Principles. Originally drawn up by Tom Beauchamp and James Childress in the context of biomedical ethics, their model is now finding use in wider contexts. As the name implies, Beauchamp and Childress (1979, and later editions) argue that a set of four key principles ought to guide any action: **non-maleficence** (as far as is possible avoid doing harm), **beneficence** (as far as is possible do good), **autonomy** (maximise the freedom of the individual(s) affected to make their own decisions) and **justice** (be as fair as possible; treat equal cases equally).

The value of principlism can to some degree be seen by the enthusiasm with which it has been adopted. Rather than slavishly following either deontological or consequentialist criteria, at its best this approach provides a balance between more rules/duties-based principles (autonomy, justice) and more outcomes-based principles (non-maleficence, beneficence).

Like any generalised scheme, of course, it is still flawed; critics argue that, by shifting the emphasis you place on each of the four principles, you can actually arrive at whatever predetermined answer you were seeking. If, for example, you were opposed to voluntary organ trading (in which someone was willing to sell one of their own kidneys), then you might emphasise the potential harm to them (either as a direct complication of the operation or in future years should their one remaining kidney fail) and to others (e.g. people being forced by family to 'volunteer' a kidney). If, on the other hand, you were in favour of voluntary organ donation you might accentuate the importance of autonomy ('the organ is theirs, if they want to sell it then that's their business') and the benefit to the organ recipient.

Another concern voiced about the four principles approach is the apparent dominance of autonomy in decision-making, and especially that a very individualistic notion of autonomy has become 'first amongst equals' (e.g. Dawson and Garrard, 2006). Whilst it is helpful to be aware of these caveats, the four principles approach nevertheless remains a useful structure upon which others, as we shall see below, have built.

9.5   Developing insight into ethical issues

It would be unreasonable and unrealistic to expect a bioscience major to become an expert in philosophy. There are, however, a series of straightforward practical steps that you can take that will equip you to have a much better grasp of key ethical and social issues associated with innovations in biology and biomedicine.

9.5.1   Turning on your bioethical radar

The first thing you can do is simply to have an expectation that bioethics is already there as part of the contemporary world. Once you get into the habit of listening out for reports of ethically relevant news, you'll discover that stories of this kind are very widely discussed in the media. Recognising the pervasiveness of bioethics is a bit like being introduced by a friend at a party to somebody else that they know but whom you initially believe you have never met. Once you've been introduced, however, you discover in subsequent days that this 'new' acquaintance is actually frequently around in the same circles as you; it was just that you were never previously attuned to their existence. Over time you speak to them on a more regular basis, and perhaps get to know them pretty well as a friend in their own right.

Getting acquainted with bioethics can be a very similar process. Once you initially become aware of an issue, you'll start to recognise familiar themes being discussed in the popular media. You can approach this familiarisation process in both a *reactive* and a *proactive* way. The reactive way would simply be to keep your bioethical radar switched on when you are reading a newspaper, watching the news on television or surfing the web. The more proactive way would be to deliberately bring bioethics stories into your world. In this day and age, there are a number of services that will deliver aggregated bioethics information directly to you. There are several very helpful websites at which you can sign up for regular e-mail and/or twitter feeds. These include Bionews www.bionews.org.uk (@bionewsUK), www.bioethics.com (@bioethicsdotcom) and my personal favourite www. bioedge.org (@bioedge). If you are RSS-savvy (see Chapter 8), then you can also set up your own feeds, or take advantage of those set up by others (e.g. the website www.bioethicsbytes. wordpress.com has a news sidebar fed by an RSS feed).

9.6   Taking it further

Having an awareness that a topic exists is, of course, only a beginning. At the risk of over-stretching my earlier analogy, it is little more than recognising that the person in the queue for the photocopier is the one your friend introduced you to at the party, but not stopping to say hello; what we want is for you to become good mates, and that, as we know from our relationships, takes both time and effort.

If you find that bioethics really starts to grab you, there is, of course, a plethora of books and journals devoted to the topic (some of which are highlighted in Section 9.11, Additional resources). One important issue you will need to consider is the extent to which you want to get to grips with the philosophical background to ethics; in other words whether you want to be an ethically interested biologist or an ethicist with an interest in biology. Some resources will anticipate that you have much firmer foundations in philosophy than is probably the case; we have tried to identify the more advanced texts within the Additional resources section.

9.6.1   New topics, new perspectives

One of the exciting, but challenging, aspects of bioethics is the regular appearance of new issues and quandaries. Technologies which were science *fiction* only a few years ago are beginning to become science *fact*, and there is no reason to imagine that this trend is going to end any time soon.

This constant emergence of fresh innovations drives home the fact that forming an opinion about any one issue, for example regarding the appropriateness of research involving embryonic stem cells, is not necessarily going to equip us to reflect upon a different issue, such as the potential to create synthetic life or the rights and wrongs of taking non-prescription medicines to improve your concentration and wakefulness during the exam season. In consequence, it is important that each of us develops some form of generic approach or framework which will provide an effective starting point for evaluating the ethics of novel developments.

At the same time, an appropriate framework will facilitate serious consideration of the views held by different stakeholders and individuals with different perspectives. It is easy to write off someone who holds a different opinion to your own on the grounds that they are *de facto* an unprincipled libertarian or a bigoted conservative (delete as applicable), but it will pay dividends in the long run to make a serious effort to understand the worldview underpinning their outlook.

It is important that the previous statement is not misconstrued. An appeal to the appropriateness of *giving a fair hearing to different viewpoints* is absolutely not to say that *all perspectives are equally accurate or equally valid*. Naive relativism of that kind can surface in a variety of different contexts. For example, some scientists, schooled in the robust testability enshrined within the Scientific Method, are dismissive of bioethics since, as they see it, ethics is merely a matter of opinions not facts. I have even heard colleagues for whom I otherwise have great admiration saying things such as 'of course there are no right answers' in relation to bioethics in a way that they would never speak of scientific experiments, even if the mechanism of a particular process had yet to be effectively elucidated.

Another arena in which there has been undue deference to the equal merit of different views has been the popular media's handling of contentious issues in science. Wary of accusations of bias, editors of news programmes have pursued a policy of strict balance between rival viewpoints, however bizarre or unorthodox they may be. In consequence, maverick individuals obsessed with a causal link between the MMR vaccine and autism, or unwilling to accept the role of HIV in AIDS, have been given equal airtime alongside scientists reflecting well-validated evidence to the contrary (Ben Goldacre's book *Bad Science* (Goldacre, 2009) and regular *Guardian* newspaper column of the same name, have been particularly candid in pointing out the inappropriateness of this editorial policy).

On the other hand, however, *appropriate* weighing up of the arguments for and against any particular development is an intrinsic part of rational reflection. Aside from the possibility that greater consideration may occasionally cause us to undertake a volte face regarding an issue, undertaking structured reflection can enable us to develop deeper evidence in support of our own position.

9.6.2   *Introducing the ethical matrix*

What framework can we use to facilitate fair reflection on a topic from a variety of standpoints? The simplest approach is to formulate an argument in support of your view on an issue and then require yourself to come up with a coherent train of thought with the opposite conclusion. An exercise of this kind certainly has value, but it also has limitations: the principal weakness being a failure to adequately reflect on the implications for other stakeholders; i.e. to give adequate consideration of the issue as viewed by someone with an opposing perspective.

It is here that the **ethical matrix** can be an invaluable tool. Devised in the 1990s by Ben Mepham, the ethical matrix is not intended to be a simple algorithm through which data can be cranked in order to come up with the 'right' ethical solution. As Mepham himself puts it, 'The matrix is designed to *facilitate*, but not determine, ethical decision-making by making explicit the relevant ethical concerns and providing a reasoned justification for any decisions made' (Mepham, 2008).

The matrix, then, is a structured way to consider the potential implications of, for example, a new biomedical innovation from the perspective of various interested parties (rows of the grid) with reference to a number of ethical principles (the columns). The principles in question are at root the four principles drawn up by Beauchamp and Childress (and discussed above). Since, however, the ethical matrix is intended to be a help not a hindrance, it is important that neither the number of stakeholders nor the number of principles becomes unwieldy. Mepham therefore recommends that no more than four interest groups are generally considered in any one matrix

Table 9.1 Ethical matrix

Stakeholder	Wellbeing	Autonomy	Fairness
Stakeholder (1)			
Stakeholder (2)			
Stakeholder (3)			
Stakeholder (4)			

and, in the other dimension, he combines two of Beauchamp and Childress' original categories 'non-maleficence' and 'beneficence' into one principle, 'Wellbeing', which, alongside 'Autonomy' and 'Fairness', provides three columns. A typical matrix would therefore consist of a four by three grid (Table 9.1). A number of worked examples of the ethical matrix in action are available online (the interactive exercises on aspects of animal farming at www.ethicalmatrix.net/ are particularly recommended).

Importantly when utilising the ethical matrix, stakeholders need not be limited to human participants; animals involved in research would clearly be included, but wider consideration of animal and plant species (possibly grouped as 'biota') may sometimes be applicable. It may not be obvious to see how the notions of wellbeing, autonomy and fairness can be applied to the broader biotic community, but, Mepham argues, these can serve as reminders of the need to factor respectively 'conservation', 'biodiversity' and 'sustainability' into the ethical evaluation process.

By utilising the ethical matrix, it should be more difficult for us either to inadvertently miss or, worse still, to deliberately dismiss consideration of a significant dimension. The latter ought to be as unacceptable as the selective reporting of only the experimental data that supports a given hypothesis.

Using the ethical matrix: before reading on, let's take another case study and see if we can get a feel for the key issues raised by the scenario through use of the ethical matrix approach.

9.6.2.1 Case study

Carl is a 21-year-old builder. He is engaged to Julie, and she has recently discovered that she is expecting their first child. In 2004, Carl's maternal grandfather (i.e. his mum's dad) died from Huntington's disease (HD), a late-onset degenerative disease of the nervous system. HD is inherited in an autosomal dominant fashion; in other words it is equally likely to affect men or women, and if you *do* have HD then you have a 50% chance of passing it on to your children. Carl's mum Maureen has decided not to take the test to find out if she got the faulty copy from her father. However, now that he is expecting to be a father himself, Carl is keen to find out if there is any risk that he has passed on the condition.

Using the ethical matrix, consider how Carl's decision about whether or not to take the test affects the wellbeing, autonomy and fairness for each of the stakeholders (Carl, Julie, Maureen and the new child).

The following matrix outlines some of the issues that may have a bearing in this case.

Stakeholder	Wellbeing	Autonomy	Fairness
Carl, the builder	If the test says Carl *does* have pre-symptomatic HD then he can plan his life to minimise the impact of the disease (e.g retraining for a job where physical strength is less important than in construction). There may be psychological issues if he tests positive. If the test says Carl does *not* have the mutant gene this may prove to be a psychological boost, but he may experience guilt if his mum (or a sibling) subsequently tests positive.	Carl is within his rights to request a test, but he may equally decide not to – his autonomy is maintained either way around.	There may be insurance implications arising from taking the test – he moves from an 'at risk' category to either being 'unaffected' or 'definitely affected'. In the latter case he may be barred from insurance.
Julie, his fiancée	Carl taking the test will remove the uncertainty that hangs over their relationship at present – there are therefore issues associated with Julie's psychological and emotional wellbeing.	Armed with the knowledge that Carl *has* the HD gene, Julie may decide she cannot go through with the marriage and/or commit to caring for Carl in later life. She may also seek a termination.	Julie may end up being keener than Carl that he has the test because it offers to clarify their situation.
Maureen, his mum	If Carl tests negative (i.e. he does *not* have the mutation causing HD), this does not prove that Maureen is also in the clear, but if he tests positive then she will almost certainly have the mutation herself and may start showing symptoms quite soon.	Maureen has elected not to have the test, so her autonomy will be infringed if Carl takes the test, and especially if he tests positive, which means she almost certainly has the faulty gene as well.	Knowing she has chosen *not* to be tested, Carl must conceal the outcome from his mum if he is to honour her wishes. Will he be able to do this once he knows his situation? Should she be actively involved in his decision?
The new baby	If Carl *does* have HD this will potentially impact on Carl and Julie's attitude towards the baby and they may even seek a termination.	At present the discussion has only been about testing *Carl*, not prenatal testing of the baby for HD. Even if Carl does have the faulty gene, there is a 50% chance that the baby does not. Knowing the result of Carl's test will not directly alter the situation for the baby. No additional testing on the baby should be performed until he/she is old enough to decide if they want the test.	The baby has inherent value as a human being and Carl's test ought to have no bearing on his/her life at this stage.

As this example demonstrates, the use of an ethical matrix has not dictated whether or not it would be right for Carl to take the Huntington's disease test. It does, however, serve to make sure that the interests of all stakeholders are clearly laid out.

9.7   Conclusion

Scientific research has a crucial role to play in contemporary society. Given both the substantial budgets involved in the research process and the dramatic impact discoveries in biology and biomedicine can have for humans, non-human species and the planet as a whole, it is essential that very serious consideration is given to the ethical conduct of researchers and the potential ramifications of their work. Students and staff alike will find their grasp of science enriched by reflection on bioethics, and it is hoped that this chapter has provided helpful pointers for the beginning of a rewarding journey into deeper understanding of these dimensions.

9.8   How you can achieve your potential in bioethics

- Watch television news and read newspapers with an active expectation that you will see accounts of controversial developments in biology and biomedicine.

- Reflect about what **you** think about the ethical and societal consequences of these developments.

- Use the ethical matrix model as a tool for reflecting on **other opinions** about a topic.

- Read journal articles that discuss the ethics of developments in biology and biomedicine.

- Read introductory books on bioethics. If you are serious about the subject, broaden your reading to include philosophical and legal perspectives.

- Form a discussion group where you meet regularly with friends to talk through the ethical implications of a development in biology or biomedicine.

9.9   Tutor notes

In all probability, most academic scientists reading this book will have had no formal training in bioethics and fewer still will have taken courses in philosophy. Bioethics as a discipline sits at the interface of a variety of more traditional fields, including sociology and economics, alongside biology and philosophy. Whereas a few years ago practitioners of 'bioethics' were likely to have started with a background in ethics and applied this to biological scenarios, there are now increasing numbers of biologists approaching the subject from the other direction.

In my experience, two 'old chestnuts' are regularly debated by staff seeking to teach bioethics to students who are first and foremost bioscientists. Firstly, how much moral philosophy should we realistically expect students of biology to take on board? Secondly, should we be assessing ethical thinking and, if so, how should we go about it? I will briefly tackle each of these issues, before adding a few additional recommendations.

9.9.1 How much philosophy ought a bioscientist to know?

As I hope the earlier text of this chapter has illustrated, students can certainly enrich their understanding of the social relevance of developments in biology and biomedicine through relatively straightforward reflection on the rationale(s) underpinning innovations. On many levels it would certainly be inappropriate for academic staff to give the impression that consideration of ethics was unimportant for the contemporary bioscientist.

We are all, nevertheless, aware of the tensions as we try to design a balanced curriculum for an undergraduate programme. The pace of research advance in bioscience, perhaps more so than any other discipline, means that the distance between cutting-edge discoveries and the background knowledge one might reasonably expect of an 18-year-old fresh from school or college becomes ever greater. There is therefore more science content to fit into a course. Alongside this there are drives to add more skill-development activities to the curriculum to enhance the employability of graduates. How much philosophical theory can we realistically add into this already crowded landscape?

My personal view is that we ought to see provision of a framework for ethical reflection on both current and, importantly, future innovation in bioscience as a fundamental aspect of any contemporary life science programme, even if some of the other molecular or organismal content needs to be omitted to make room. We cannot, and should not, expect to turn bioscience majors into professional philosophers, but analysis, at least as deep as that covered in this chapter ought to be feasible for any course. If you have sufficient flexibility within the curriculum and access to the necessary expertise (perhaps via involvement of colleagues from a philosophy, sociology or theology faculty on campus), it may be possible to introduce both an introduction to bioethics within a core module and deeper analysis in an optional unit.

9.9.2 How should we assess ethics?

A second concern for bioscience staff relates to assessment of ethical activities. Scientists tend to be more familiar, in general, with awarding credit for re-presentation of factual content, and marking of ethical thinking can place them outside their comfort zone. Faced with this dilemma, one route is essentially to 'cheat' and base the assessment on more familiar criteria such as accuracy of science within the content and on presentation skills. We ought, however, to get beyond this and ensure that marks are being given for genuine engagement with the ethical process.

Here I think the QAA benchmarking statements, noted at the beginning of the chapter, are helpful. In particular the aspiration that a good honours graduate ought to 'be able to construct reasoned arguments to support their position on the ethical and social impact of advances in the biosciences'. Marks can be awarded for the clarity of logical thought and reasoning (even, importantly, if the student's conclusion differs from our own view, provided they have offered appropriate justifications). Credit can also be given for serious consideration of an issue from more than one perspective and for the extent of engagement with core ethical philosophies.

9.9.3 How should we introduce bioethical topics?

As the earlier scenario involving Craig and Jennifer Wyatt (and their potential involvement in a trial using fetal eggs for IVF) was intended to illustrate, case studies can serve as excellent vehicles for opening up complex issues. Approaching bioethics from the philosophical 'end' with a class whose primary interest is bioscience is unlikely to engage the majority of the group; case studies demonstrate the relevance of ethical thinking and allow real-world framing of otherwise abstract ideas.

Choice of cases to discuss is important; they can be genuine or fictional, though the latter work best when as close to real situations as possible. Real cases arise naturally and frequently in news stories; these accentuate the relevance of bioethics, but the fact that they relate to the lives of real people means that they may involve specific details that add unnecessary complexity. Fictionalised versions of cases can ensure that the key issues surface more naturally as a result of details you have omitted and/or added to the scenario.

The degree to which details and twists are introduced into your case studies may depend upon the ways you intend to use them. For example, interaction with a scenario might be limited to 'Here's a short report from last night's news. Having watched the clip, turn to your neighbour and discuss whether you would allow this treatment to occur and why.' At the other end of the spectrum, you might use the case to launch a full-blown role-playing activity lasting for an hour or more. Alternatively, I have seen effective engagement achieved by asking participants to act as the ethics committee deciding if a development ought to be permitted (Mark Goodwin, personal communication) and via a more rapid-fire 'doughnut debate' (Macer, 2008). During the latter, two rings of chairs are set out, with the occupants facing each other, such that two individuals can have a one-to-one conversation. Students sitting in one of the rings, e.g. the one facing outwards, can be asked to argue for one minute in favour of the development posed in the scenario (regardless of their personal beliefs on the matter). The student facing them can then be asked to offer evidence in support of the alternative viewpoint. This approach has a number of attractions: it requires all members of the group to become engaged in the task, but in a relatively non-threatening manner; it forces students to think about other perspectives on a topic; and the circular layout makes it easy to shuffle the groups for a further activity by asking members sitting in one of the rings to move one seat to their right.

9.10   References

Arthur, C. (2003) Early death of Dolly the sheep sparks warning on cloning. *Independent* 15th February 2003. Available at: www.independent.co.uk/news/science/early-death-of-dolly-the-sheep-sparks-warning-on-cloning-597652.html. Accessed April 2011.

Bae, H.-J. (2009) Stem cell scientist found partially guilty. *Korea Herald* 27th October 2009. Available at: www.koreaherald.com/national/Detail.jsp?newsMLId=20091027000077. Accessed April 2011.

Beauchamp, T.L. and Childress, J.F. (1979) *Principles of Biomedical Ethics*. Oxford: Oxford University Press.

Black, R. (2010) Nature loss 'to damage economies'. BBC News website 10th May 2010 (www.bbc.co.uk/news/10103179). Accessed April 2011.

Briggs, H. (2001) An oath for scientists? BBC News website 30th March 2001 (http://news.bbc.co.uk/1/hi/sci/tech/1250331.stm). Accessed April 2011.

Bryant, J., Baggott la Velle, L. and Searle, J. (2005) *Introduction to Bioethics*. Chichester: John Wiley & Sons, Ltd.

Butler, D. (2010) Journals step up plagiarism policing. *Nature* **466**, 167.

Cameron, N.M. de S. (2004) Christian vision for the biotech century. In *Human Dignity and the Biotech Century*, Ed. Colson, C.W. and Cameron, N. M. de S. Downers Grove, IL: InterVarsity Press.

Coady, C.A.J. (2009) Playing God. In *Human Enhancement*, Ed. Savulescu, J. and Bostrom, N. Oxford: Oxford University Press.

Curry, P. (2006) *Ecological Ethics: An introduction*. Cambridge: Polity Press.

Dawson, A. and Garrard, E. (2006) In defence of moral imperialism: four equal and universal prima facie principles. *Journal of Medical Ethics* **32**, 200–204.

DIUS (2007) *Rigour, Respect, Responsibility: A Universal Ethical Code for Scientists*. London: Department for Innovation, Universities & Skills (DIUS). Available at www.berr.gov.uk/files/file41318.pdf. Accessed April 2011.

Dyer, O. (2008) Psychiatrist admits plagiarism but denies dishonesty. *British Medical Journal* **336**, 1394–1395.

Eriksson, S., Helgesson, G. and Höglund, A.T. (2007) Being, doing, and knowing: developing ethical competence in health care. *Journal of Academic Ethics* **5**, 207–216.

Ewing, A.C. (1953) *Ethics*. London: English Universities Press.

Gibson, I., Dhanda, P., Harris, T. *et al.* (2003) *House of Commons Science and Technology Committee: The Scientific Response to Terrorism. Eighth Report of Session 2002–03*. Volume I. London: The Stationery Office. Available at: www.publications.parliament.uk/pa/cm200203/cmselect/cmsctech/415/415.pdf. Accessed April 2011.

Goldacre, B. (2009) *Bad Science*. London: Harper Perennial.

Grenz, S.J. (1998) *The Moral Quest*. Leicester: InterVarsity Press.

Harris, J. (2001) *Bioethics*. Oxford: Oxford University Press.

Hickman, L. and Randerson, J. (2009) Climate sceptics claim leaked emails are evidence of collusion among scientists. *Guardian* 20th November 2009. Available at: www.guardian.co.uk/environment/2009/nov/20/climate-sceptics-hackers-leaked-emails. Accessed April 2011.

Holland, S. (2003) *Bioethics: A Philosophical Introduction*. Cambridge: Polity Press.

Hutchinson, M. (2003) Outrage over aborted eggs plan. BBC News website 1st July 2003 (http://news.bbc.co.uk/1/hi/health/3034266.stm). Accessed April 2011.

Macer, D.R.J. (2008) *Moral Games for Teaching Bioethics*. Haifa, Israel: UNESCO Chair in Bioethics. Available at: http://unesdoc.unesco.org/images/0016/001627/162737e.pdf. Accessed April 2011.

Mepham, B. (2008) *Bioethics: An Introduction for the Biosciences* (2nd Edition). Oxford: Oxford University Press.

Potter, V.R. (1971) *Bioethics, a Bridge to the Future*. Englewood Cliffs, NJ: Prentice-Hall.

QAA (2007) *Biosciences: Subject Benchmarking Statements*. Gloucester: Quality Assurance Agency for Higher Education. Available at: www.qaa.ac.uk/academicinfrastructure/benchmark/honours/biosciences.pdf. Accessed April 2011.

QCA (2007) *Science: Programme of Study for Key Stage 3 and Attainment Targets*. London: Qualifications and Curriculum Authority. Available at: http://curriculum.qca.org.uk/uploads/QCA-07-3344-p_Science_KS3_tcm8-413.pdf. Accessed April 2011.

RCUK (2009) *RCUK Policy and Code of Conduct on the Governance of Good Research Conduct: Integrity, Clarity and Good Management*. Swindon: Research Councils UK. Available at: www.rcuk.ac.uk/documents/reviews/grc/GoodResearchConductCode.pdf. Accessed April 2011.

Shamoo, A.E. and Resnik, D.B. (2009) *Responsible Conduct of Research* (2nd Edition). New York: Oxford University Press.

Simberloff, D. and Stiling, P. (1996) How risky is biological control? *Ecology* **77**, 1965–1974.

Thomasma, D.C. (1997) Bioethics and international human rights. *Journal of Law, Medicine and Ethics* **25**, 295–306.

Willmott, C. and Willis, D. (2008) The increasing significance of ethics in the bioscience curriculum. *Journal of Biological Education* **42**, 99–102.

Wilsdon, J. and Willis, R. (2004) *See-Through Science: Why Public Engagement Needs to Move Upstream*. London: Demos. Available at: www.demos.co.uk/files/Seethroughsciencefinal.pdf. Accessed April 2011.

Wilson, E. (2002) Apology. *Trends in Biochemical Sciences* **27**, 64.

Yeoman, F. (2008) Twins who were separated at birth 'in accidental marriage'. *Times* 11th January 2008. Available at: www.timesonline.co.uk/tol/news/uk/article3171716.ece. Accessed April 2011.

9.11 Additional resources

As mentioned previously, books on bioethics by Mepham (2008) and Bryant *et al.* (2005) are intentionally written with ethically interested bioscientists in mind, and are an excellent next step for biologists wanting to read more widely on the ethical dimensions. As far as journals and academic magazines are concerned, the *Journal of Medical Ethics*, the *American Journal of Bioethics* and the *Hastings Center Report* are amongst the most accessible in terms of content (though, sadly, may require a personal or institutional subscription in order to read them). Other general science periodicals such as *Nature, Science*, the *British Medical Journal* and *New Scientist* frequently discuss the ethical and societal significance of recent innovations. Provided you bear in mind the usual caveats about the uncertain authority and expertise of the writer, *Wikipedia* can sometimes include quite detailed discussion.

Some resources have a higher expectation of your background in philosophy. For example an anthology of readings by several major authors (e.g. Harris, 2001) or single-author volume (e.g. Holland, 2003) will reward the persistent student, but would not be a good place to start your deliberations.

A special mention for reports by the Nuffield Council on Bioethics: in terms of materials available via the Internet, I'd like to pick out one website for particular mention. Since 1991, the Nuffield Council on Bioethics have, on average, produced one substantial report per year on a bioethical issue. These reports are thorough (often 200 pages or more in length) but accessible to the non-specialist (there's also a user-friendly summary version of each report). The Nuffield reports are all available as free downloads from their website, www. nuffieldbioethics.org.

The UK Centre for Bioscience's *Ethics in the Biosciences Briefing* brings together references and resources recommended by the bioscience learning and teaching community. The briefing will be of use and interest both to those with experience and those new to the topic. Available at www.bioscience.heacademy.ac.uk/resources/briefings/ethics.aspx.

10 Assessment, feedback and review

Stephen J. Maw and Paul Orsmond

10.1 Introduction and some definitions

The word 'assessment' conjures up a range of images in peoples' minds; sadly many of these are negative. However there is much more to assessment than regurgitating information for written examinations, and modern teaching practice recognises that assessment is central to learning. Your university will use assessment to do the following:

- decide whether you can progress to the next stage of your studies or to graduation;

- classify your performance and 'rank' it alongside the performance of other students;

- guide and improve your learning.

In this chapter we describe how you can make the most of the whole process of assessment during your time at university.

At this point it is useful to introduce some of the terms you will encounter from time to time. There are two types of assessment: summative and formative. A *summative assessment* is one that 'counts', i.e. you will be awarded a mark or grade which contributes to the final mark for the course or module. Often, but not necessarily, summative assessment occurs at the end of a module. You are probably less familiar with the process of *formative assessment*. In this case, while you may be given a mark for the work, it will not 'count' in the sense that it will not contribute to the final module or course mark. However, that is not to say it is unimportant or can be ignored. Whatever you are learning it is crucial to practise and give things a go – formative assessment allows you this opportunity. It enables both you and your tutor to see how well you are doing and where there may be room for improvement. Whichever type of assessment is used, one way in which tutors can guide you in tackling an assignment is through discussion of the *marking criteria*. For each criterion your tutor will use a range of performance *standards*. Let's take an analogy from athletics and the criterion of 'finish time' for a marathon. If you are an elite male athlete then an excellent finish time would be

Effective Learning in the Life Sciences: How Students Can Achieve Their Full Potential, First Edition.
Edited by David J. Adams.
© 2011 John Wiley & Sons, Ltd. Published 2011 by John Wiley & Sons, Ltd.

2 hours 20 minutes; if you are a club runner you may hope to finish in less than 3 hours, and for most of the population just finishing is a great achievement! Performance for all who take part can be measured against these and other standards.

There are two types of marking criteria: a generic form, where the tutor is looking for general abilities (for example, clear referencing of literature cited) and more subject-focused criteria which involve assessment of specific aspects of the assignment (for example, correct use of disassociation constants in biochemistry). Marking criteria are closely related to module *learning outcomes*. These describe the knowledge and skills you should have acquired after completing the module/course. It is clearly essential that you understand both the learning outcomes and marking criteria for a module; the latter will be used to assess whether or not you have achieved the former.

Whenever your tutor returns assessed work it should have some *feedback* associated with it. Tutors can provide feedback in many different ways; for example: discussion during a tutorial or even in the corridor; written corrections or notes (annotations) on a piece of work; an email to you or the whole class; a five-minute talk at the beginning of a lecture; it may even be a video recording. Irrespective of the format, feedback should not just be a commentary on what you have done and how well you have achieved the learning outcomes: it should also provide pointers on how to improve next time – this advice is often termed *feed-forward*.

Until recently, all assessments at university would have been marked by a member of staff or postgraduate student. However, research has shown that students can benefit greatly from assessment procedures that involve evaluation of their performance by other students: so-called *peer assessment* or *peer review*. It is likely that, from time to time during your degree programme, you will be asked to evaluate other students' work and have your work assessed by your classmates. Although you may feel uncomfortable about this at first, you are likely to find that it provides a very powerful learning experience. Likewise you may be asked to critically evaluate your own performance – *self-assessment*. The ability to evaluate your own performance and that of others is a skill highly regarded by employers.

Finally, the troublesome issue of *plagiarism* – something that must be avoided at all costs. Plagiarism may be defined as copying someone else's work and presenting it as your own. Universities impose **very** stiff penalties for plagiarism and you should read very carefully the guidelines your university provides on how to avoid plagiarism. Plagiarism in the context of e-learning, and ethical issues associated with plagiarism, are considered in Chapters 8 and 9, respectively.

Should you wish to know more about assessment and feedback, and the various terms introduced and defined in this section, you should find the student short guides *Assessment* and *Feedback – Making it Work*, produced by the UK Centre for Bioscience, of value (see Section 10.12, Additional resources).

10.1.1 *Tutor notes*

John Cowan, former director of the Open University in Scotland, famously describes assessment as the engine that drives learning; yet speak to many staff about marking and their faces fall. Further, every year since the inception of the National Student Survey in 2005, the results have told us that assessment and feedback continue to be the least satisfactory element of the student experience. As David Boud indicates, 'students can escape bad teaching; they can't avoid bad assessment' (Brown and Smith, 1997). According to Phil Race (2009), 'we need to reduce the overall burden of assessment for ourselves and for our students. We need to measure less, but measure it better.' For those wishing to improve their assessment practice, the problem is not a lack of literature but where to start. One good place is the UK Centre for Bioscience's *Assessment Briefing* (see Additional resources), which provides information on resources and tools for assessment in the biosciences; it is written with the needs of the lecturer in mind. Equally the Centre's *Assessment Audit Tool* (see Additional resources) provides a good way for departments to initiate discussion and development in this area. The Tool makes clear the range of activities involved in delivering assessment and can identify both departmental strengths and areas for improvement.

10.2 Types of assessment

Nowadays, bioscience tutors are likely to utilise a wide range of formal, summative assessment procedures. These include:

- multiple-choice questions (MCQs)

- short-answer questions (SAQs)

- essays, including essay-style exam questions

- practical skills tests (laboratory or field)

- laboratory or fieldwork reports

- oral presentations

- poster presentations

- design of web pages

- problem solving exercises

- data interpretation exercises

- literature search exercises

- dissertation.

You should find out precisely how each module will be assessed at the beginning of the course. For example, there may be four assessments during the module (perhaps a MCQ test, an essay, a poster presentation and a data presentation exercise) that will each be worth 10% of the final mark for the module, and an end of module exam worth 60% of the final mark.

A great advantage of using a variety of assessment procedures is that together they enable tutors to measure a wide range of abilities. For example, MCQs can be used to assess the extent of a student's general knowledge of a bioscience subject, practical assessments give an indication of manipulative skills in the laboratory or field, and assessed oral presentations give an idea of a student's ability to communicate results and ideas effectively. A key point here is that you will be given a score for each summative assessment procedure, and the accumulated scores will then be used to give you an overall mark for the module. This mark will, in turn, determine whether you can progress to the next stage of the course and, in the later years of the programme, is likely to contribute to your degree classification (whether you obtain a first class, 2.1, 2.2 degree etc.). It is therefore essential that you fully understand how tutors will mark each assignment; i.e. you should consult the *marking criteria* for each piece of work.

10.3   Marking criteria

A formal definition of a criterion is a standard by which something may be judged or decided. When a tutor marks your work it is likely that he or she will refer to several marking criteria before awarding a final score. It is essential that you should be fully aware of the marking criteria, and you should make sure you understand these criteria **before** you begin an assignment. As an example, the marking criteria for the assessment of poster presentations could be considered under the following headings:

Coverage and relevance
Has the author focused on the topic? Is there good breadth of coverage and integration of key concepts?

Analysis
Is it clear that the author understands the material and does he/she explain concepts clearly? Is there evidence of critical analysis of the information presented? Are conclusions justified?

Supporting evidence
Is there good evidence of extensive and critical reading around the subject? Are key points illustrated with well-chosen examples from the literature?

Presentation – visual
Does the poster have visual impact? Is the layout appropriate – are individual sections clearly defined? Is there a good balance of text and diagrams?

Presentation – oral
Is verbal delivery clear, well paced, with variability in tone and expression, and of suitable volume for the environment?

Your university should publish all marking criteria, along with the learning outcomes for each module, in readily accessible locations (usually in printed or online module or programme manuals).

10.4   Learning outcomes

The learning outcomes summarise the knowledge, understanding and skills that you, the learner, can be expected to acquire during a module or programme. As with marking criteria, you should make sure that you can access information about learning outcomes for all of the modules you will take. Here are some of the intended learning outcomes given in the module handbook for a first-year module 'Introducing Biomedical Science' at a UK university:
By the end of the module students should be able to:

- *demonstrate awareness of the key features of an effective essay-style examination answer;*

- *show how to construct a report to convey the results of a scientific investigation;*

- *discuss the basis of scientific investigation in the context of analysing the behaviour of macromolecules such as proteins and nucleic acids;*

- *carry out basic numerical manipulations of scientific measurements, and critically analyse experimental data;*

- *demonstrate knowledge about how to prepare a scientific poster;*

- *offer constructive advice to their peers;*

- *accept advice and feedback from tutors and/or peers and focus these into suitable targets for personal development.*

The learning outcomes were accompanied by the following information:

> *There are four pieces of assessed work within this module, each worth 25% of the mark for this course. These are: (1) an exam-style essay; (2) a poster presentation; (3) an experimental report; and (4) a data handling assessment.*

Learning outcomes allow you to focus and decide where you ought to concentrate your efforts. Therefore it is essential that you fully understand what they mean and the extent of the skills and knowledge you will be expected to acquire during the course of your studies. For example 'demonstrate knowledge about how to prepare a scientific poster' is clearly going to be assessed in the second assessment; the criteria listed in Section 10.3 indicate the various aspects of the poster and its presentation that might be evaluated.

10.5 Feedback

10.5.1 *Getting ready, understanding and using feedback*

During your studies you will get lots of feedback, and to make the most of it you need first of all to recognise it and then be receptive to it (Race 2008). In its broadest sense, feedback is when you receive comments on your work that tell you how well (or badly) your studies are going. A good way to look at feedback is to think of it as useful advice. Most feedback is likely to come from your tutor, but it may also come from module leaders, lab demonstrators and fellow students. The ASKe CETL (see Additional resources) suggests three steps to help you get the most out of feedback:

(1) get ready for feedback

(2) understand feedback

(3) use your feedback.

The first step is not difficult or intellectually challenging; it is largely about organisation: you need to look at the module handbook or ask your tutor so you know beforehand when and where the feedback will happen, and in what form. For example, if, following marking of essays, a feedback session on the correct way to construct an essay is to be held during a tutorial, you must make sure you attend and benefit.

You need to make sure you understand feedback by reading, watching or listening to it carefully and considering how it applies to your work. A key aspect of this is how the feedback relates to the assessment criteria. It may seem obvious, but you cannot make good use of feedback if you don't understand it fully or cannot read it. Feedback is primarily about a dialogue between you and the tutor; so if you don't understand then be prepared to ask.

Finally, how are you going to use the feedback? Good feedback will do more than explain why you got a certain grade (assuming you got one); it will probably point out errors you made but it will also make suggestions on how you could improve in the future (feed-forward). To be successful you need to find a way to remember this advice and apply it in the future. You may wish to divide the points into major and minor. Minor points might include punctuation errors or failing to write a species name in italics. Major points could include things such as incorrect referencing, failure to calculate a concentration correctly or not knowing what is meant by a one-molar solution – things that you will need to spend some time researching before your next assignment.

10.5.2   The feedback loop

One way to get the best out of the feedback you receive is to develop an action plan that lists the key points raised by the feedback and how you intend to respond to these. For example, you might make a note to consult your tutor about a specific issue, read more widely around a particular topic or practise a skill. Give yourself a deadline to meet these targets. One of the most common mistakes students make is to restrict their use of feedback solely to the piece of work for which it was given. When considering feedback comments, think about how you could use the advice you have received more widely.

The ASSET Project investigated ways in which staff may use video clips to give generic (i.e. general) feedback to their students (Figure 10.1).

Although the diagram is intended as a guide for staff, it does illustrate some important points; the most obvious being the cyclical nature of the learning process and the fact that feedback is only part of that process. You will note that the sixth stage of the cycle feeds back to the first. This is perhaps the aspect of feedback most commonly overlooked by students – it is important

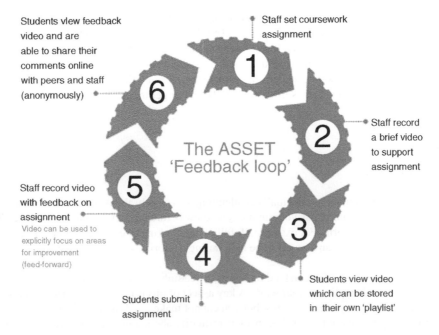

Figure 10.1   The ASSET video feedback loop

that you think about how the feedback you receive for a particular assignment feeds forward into (i.e. informs) subsequent work in the same and different modules. Another useful point to note from Figure 10.1 is that the ASSET project provided an opportunity for students to share their comments with staff and other students. Social elements of learning are important and are explored further in Section 10.6.

10.5.3 Ignore the mark

We said earlier in the chapter that it is important to be receptive to the feedback you receive. Race (2008) discusses this in detail, but essentially there are two issues that can easily knock you off track: the mark you receive, and how you handle criticism or praise. One point that staff often make is that 'students are only interested in the mark'. Now if you receive back, let's say, an essay with a mark and some written comments, it is only human nature to look at the mark first. After all, the mark is a form of feedback, albeit a limited one. It is easy to compare your mark with the mark obtained by your friends and to link it to degree classification. If the grade is what you hoped for then there's a strong temptation to ignore the written comments or give them a superficial read through – after all you have achieved what you wanted. However, in doing this you will waste vital information that could help you improve in future tasks. For example, an overall mark of 60% may mean that all elements of the assignment were completed to that standard. Alternatively, you may have completed two components very well and a third very badly. It would be a waste of good feedback not to make use of any pointers you received on how to improve weaker aspects of your performance in future.

The mark you obtain may be well below what you'd hoped for or expected, and it is easy to become defensive, dismissive or defeatist. Here are Frank's comments to his housemate after receiving a low mark: 'I just can't understand why I got a low mark for this essay – what's the point – I hate ecology. I should never have taken this module – I'm just crap at writing essays.' Clearly there is a lot of emotion expressed in these statements, and when you get a low mark there is a danger that emotions can take over. If that happens the channels of communication break down and feedback is no longer the dialogue it is supposed to be. In such circumstances it is perhaps best to give yourself some time and return to the feedback at a later date when you can be more objective. Frank's comments contain some interesting viewpoints: 'I'm just crap at writing essays' is a very defeatist attitude and not helpful. His statement may be entirely unjustified: he may have obtained a low mark because he misunderstood the question; i.e. he may have produced a beautifully written, well-structured, grammatically correct essay that happened to be on the wrong subject! It is not possible to tell very much from an overall mark; the student needs to read the tutor's comments. On the other hand, it may be that the essay was poor, but that does not mean Frank is incapable of writing a good essay. Some people are better than others at writing essays, but we can all improve with practise and guidance.

If the mark you obtain for an essay is better than you expected then there is the danger you may feel smug or rest on your laurels. Equally, some people find it hard to handle praise and are inclined to play down a compliment – 'oh it's not that good'. Say that often enough and you will believe it, and in doing so distort your understanding of the true value of your work.

Given the way a mark can influence people's reaction to, and interaction with, feedback, it is perhaps not surprising that educators advise students to look first at the feedback, decide on the mark they think they are likely to have been awarded and only then look at the actual mark awarded by the tutor. Such an approach requires discipline but can reap large rewards. Your tutor may actually help you to do this by returning pieces of work without a mark, or initially withholding the mark, to allow you time to rationally consider the feedback.

If you wish to explore further your feedback preferences, and gain a deeper understanding of how you react to different forms of feedback, you should find *Building on Feedback* (Race, 2008) useful. It contains further information and tasks to help you better understand how you react to the marks you are awarded and the comments you receive from tutors. It should therefore help you to see how you can make more of feedback.

10.5.4 Types of feedback

Feedback frequently involves the provision of tailored, written comments relating to a specific piece of work. However, just as the types of assessment used in university degree programmes are more varied than they used to be, so, particularly thanks to improvements in technology, are the ways you may receive feedback. Type-written feedback removes the issue of illegibility, but oral and visual feedback open up richer possibilities as they can convey messages through tone, intonation and body language. If you are given oral or visual feedback in electronic form (e.g. as an mp3 file) then you have the opportunity to re-listen or re-watch and check any points. If you are given the information 'live' (e.g. in a tutorial) then good advice is to take detailed notes. Human memory can be very selective: during feedback you may receive an equal number of positive and negative comments but after a while you will remember more of the negative ones. Giving feedback is very time consuming for tutors, especially if there is the added step of recording a video, so to make this manageable they often give generic (i.e. general) feedback. For example, during the first five minutes of a lecture your tutor may go through some of the common mistakes made in a recent practical write-up. You should take notes then later sift through the information, identify comments relevant to you and make sure you act on the advice given.

While feedback is usually associated primarily with marked coursework, many universities now offer 'exam surgeries' at the end of each semester. These provide an excellent opportunity for you to look at marked exam scripts and take careful note of the tutor's annotations/comments relating to your performance. You should make full use of any exam surgeries offered by your university.

10.5.5 Tutor notes

Feedback is essential for effective learning. Bellon *et al.* (1991) state 'academic feedback is more strongly and consistently related to achievement than any other teaching behaviour . . . this relationship is consistent regardless of grade, socioeconomic status, race, or school setting.' There is no ideal feedback method which can be standardised, but all feedback needs to be relevant, timely and in a language students can understand. Above all, feedback is about dialogue with the student and allowing space to check comprehension, discuss ideas and explore how to move forward.

Often feedback is given in a written form, but now the use of audio and video feedback are becoming popular and have been shown to be effective. Merry and Orsmond (2008) showed that effectiveness in audio feedback depended in part on the students recognising variation in the tone and intonation of the voice. Lunt and Curran (2011) showed that audio recording helped overcome some of the key student concerns, including late return and quality of feedback, identified in an NUS survey. Northcliffe and Middleton (2008) carried out an interesting study called 'Walk Through', where they recorded their discussion with a student as they provided feedback on a one-to-one basis. Each student took the recording away and played it later:

listening to the feedback helped them relive the original experience, and they indicated that they heard things as if for the first time.

Video recordings of feedback are becoming increasingly popular. The ASSET project (see Additional resources) investigated the use of video media to provide feedback and feed-forward to students, and is a good source of information for those using, or looking to use, video as a means of generic feedback.

Any realistic consideration of feedback needs to address the vexed issue of staff time and workload. Race (2005) examined the efficiency of different feedback methods during learning and showed that certain feedback techniques are far more effective than others. Glover and Brown (2006) dissected the written feedback given to students at two institutions; their recommendations were: to avoid trying to correct all the errors students make; to focus instead on the main weaknesses; and to give feedback to help students become better at error detection. In autumn 2002, the Centre for Bioscience published a themed edition of their *Bulletin* on feedback and feed-forward (see Additional resources). The majority of the articles were written by practising bioscientists, and as such describe realistic approaches to providing feedback. Similarly, the *Engage in Feedback* website (see Additional resources) has links to interviews with bioscience teaching staff on their views and experience of giving feedback to students.

10.6 Peer support – learning from and with your classmates

A social approach to learning is of great value – you and your peers can do a great deal to help one another make the most of your time at university. There follows a typical discussion involving three students (peers). You have probably had similar discussions at school or university.

10.6.1 Case study

Jen, Alice and Jules each prepared a 2000-word essay on mechanisms underlying the sense of taste. They discussed their approach and progress with the essay at three stages: before they started work; two weeks later; then, finally, four weeks later just before the essay was due to be submitted to the tutor.

At the start

JEN: '2000 words – where do I start?'

ALICE: 'I know, but it has to be done. We need a plan. I've got some books out of the library which may help.'

JULES: 'First we have to look at what we are supposed to be doing. What's the title?'

ALICE: 'It's "In preparing food we often apply a little salt for taste. Discuss the mechanisms underlying the sense of taste." It is not too bad: taste buds and that sort of thing.'

JEN: 'That is just so boring.'

JULES: 'OK so let us make a plan. What are the key words?'

JEN: 'Salt, taste – that's OK I did salt at school, that tongue map thing – discuss . . .'

ALICE: 'Let me draw a mind map; it just helps me think.'

Two weeks later

JEN: 'I've found out about this disease where the person can't taste salt; some genetic stuff is missing in the taste bud. Something in the receptor doesn't work.'

JULES: 'Yeah I found that too; one of the alpha subunits doesn't function. I've also got a load of stuff on nerve impulses to the brain; so that should help.'

ALICE: 'OK, so what do we have? (Alice draws a mind map.) Salt . . . dissolves in the mouth, gets taken into the receptor and what?'

JEN: 'Causes the nerve to fire and send a message to the brain and . . .'

JULES: 'Well that's OK, but what causes the nerve to fire; what is the link between the receptor and the nerve?'

ALICE: 'Where is the plan we did at the start? We had something about that . . . OK here it is . . . yes, it's the bit on depolarisation and neurotransmitters being released.'

JEN: 'Yes, that's the problem with this disease, there is no depolarisation – whatever that is. Alice I think these maps and plans are really good. Can you show me how you do them?'

ALICE: 'Yeah.'

JULES: 'OK, so we need stuff on the depolarisation and . . .'

ALICE: 'And I have this. It's in the magazine called *Science*, and it's about how the taste map of the tongue is all wrong and that nerves don't just fire to one thing like salt-specific, but there is some multiple firing pattern.'

JEN: 'What is a multiple firing pattern? We learnt about that map of the tongue in school – it can't be wrong.'

JULES: 'Can I borrow that article? The idea of multiple firing patterns helps me understand the tongue better. I never understood the tongue maps; it just didn't seem to make sense.'

JEN: 'Why doesn't it?'

JULES: 'Well how could only certain parts of the tongue respond to certain chemicals – I don't just taste salt on the left side of my tongue, do I?'

JEN: 'Yeah, that sort of makes sense, but at school they told us about the tongue map. Ahh I don't know; both seem right. I hate this; why can't we just be told?'

Four weeks later

ALICE: 'Right, I've almost finished. I've just got the end bit to do. I had to do a bit of a rewrite last night as I read it and it didn't make sense.'

JEN: 'I need to re-read stuff more often; I'm always getting feedback saying "read your work before you hand it in", but I still haven't.'

JULES: 'What about that marking criteria; have we met that? You know how David [tutor] is always banging on about that.'

JEN: 'I never use that – I just can't understand it.'

JULES: 'It can help. Look, you just look at what you need to include in the essay and it tells you how you can get extra marks by doing certain things.'

JEN: 'OK, so why didn't someone tell me that at the start!'

ALICE: 'Thing is, perhaps we should have used it at the start; perhaps this should have been part of the plan. Next time we do an assignment we should start with these criteria and these learning outcomes. I mean looking at them both now, these outcomes are really helpful. Those books I got at the start, well only some were useful. If I'd used these outcomes, I could have focused my search.'

JEN: 'I thought outcomes and criteria were the same thing . . . it is so confusing . . . I hate essays.'

JULES: 'Alice, that is a good point; I could have used these outcomes to better explain nerve depolarisation. If I were doing this again I think maybe I'd have written this differently.'

Exercise

Take some time to write down, or discuss with another person, how you think the students may have helped one another as they planned and completed their essays.
Here are some of the important points we identified from the discussion.

(1) The conversation was in a language and style that everyone easily understood.

(2) Each student made a distinct contribution to the discussions. For example, Alice helped Jen adopt a 'mind map' approach (see Chapter 1) to the exercise; while Jules made Alice and Jen more aware of the importance of the learning outcomes and marking criteria for the assignment.

(3) Together the students tried to make sense of things and helped one another learn. Jules described the conflict between what she had read and her experience of taste, but indicated that the idea of multiple firing patterns now made sense to her. The peer discussion helped her construct a more meaningful understanding of taste.

(4) The students made one another aware of new areas to explore. For example, Alice told Jules about a highly relevant article in the journal *Science*.

(5) Jen repeatedly showed signs of wanting to be told what to do rather than form her own opinions. If you are to progress at university you will need to think for yourself and show initiative. In a friendly environment Jen's classmates encouraged her to show more independence.

(6) It seems clear that interaction with peers made the students think about how they learn and the way they should approach an assignment. All three students indicated that they would be likely to do things differently next time.

The conversation between Jen, Alice and Jules suggests that classmates can do a great deal to help one another learn. In addition, as they review the performance of their peers (and themselves) in a friendly and supportive environment, they develop their capacity to assess work fairly and objectively. Tutors understand the value of peer review and will help you to further develop your assessment and evaluation skills during more formal peer assessment exercises.

10.7 Peer assessment

At some time during your studies you are likely to be asked to evaluate your classmates' work. For example, you may be required to mark a practical write-up submitted to your tutor by one of your peers. Peer assessment normally involves the whole class and takes place, under supervision, in a lecture theatre or laboratory classroom. You should be given the correct answers, along with clear guidance on the marking procedure, and asked to award marks to the various components of the write-up.

Depending on your previous educational experience you may or may not be familiar with the concept of peer assessment. If this is the first you have heard of it, you might be feeling a little apprehensive about the whole process; you may even feel that it's not your job; rather it's something your tutor is paid to do. If you feel this way you should know, first of all, that there is good correlation between marks awarded by tutors and those awarded by peers. Furthermore, there is good evidence that students can benefit greatly from participation in peer review exercises. Here are some of the things you might get out of this method of assessment (adapted from Hughes, 2001):

- The marking system is entirely open to you; this means you will see exactly what is required by markers.

- You will get a better understanding of how the standard of your work compares to the work of others.

- If you are to benefit maximally from peer assessment, your tutor should provide a full explanation and justification of the marks you will award. This should greatly improve your understanding of the science involved.

- In assessing others you will develop the ability to stand back from your own work and assess that as well. This is an essential skill in a scientist: the capacity to give an unbiased and objective assessment of the standards *you* have achieved in your own work.

- You will develop the ability to assess others fairly and objectively. This will involve coming to terms with the problem of bias; someone who is a good friend may have submitted poor work and it can be disturbing to have to give them a low mark.

Your involvement as a peer assessor can help demystify the whole process of assessment, as you use the tools and approaches adopted routinely by your tutor. Peer assessment can also be seen as providing an 'apprenticeship' in judgement and decision making; the objective assessment skills you acquire will prove invaluable in your future career as a bioscientist or in other workplace settings.

10.7.1 Case study

Scott and his classmates were shown five posters, generated by students from previous years, that illustrated an important topic in microbiology. They were asked to write a paragraph on the sort of feedback they would provide to the authors of the posters.

Scott was asked whether he found the process helpful – here is part of his reply:

> *Just looking at one poster, it is hard to think what I can write; it's difficult to know what to say. But when you have seen three or four you have a much better idea of what to write. Looking at somebody else's work, you get ideas from that piece of work. You think: 'I never thought of doing that.' Or you may see things, get ideas that you have not taken into consideration.*

Even with only minimal guidance Scott found the process beneficial. He was able to see variation between the poster presentations and picked up some ideas that he can see himself using in the future. However, he would have been able to make fairer and more objective judgements if he had framed his evaluations around some marking criteria.

Exercise

One way to further develop **your** assessment skills is to learn how to evaluate work using agreed marking criteria. It's likely that many scientific posters adorn the walls of the Bioscience Faculty at your university. Have a look at one or more of these; consult the marking criteria for posters listed in Section 10.3, then decide on the percentage mark you would award each poster(s). You should assume that each of the marking criteria carries equal weighting. Think about the feedback you would provide to the author(s) of the poster(s).

The exercise in Section 10.7.1 should allow you to practice and further develop your critical, evaluative skills. These will be further enhanced by you learning how to generate marking criteria of your own. As you develop these skills you will acquire the ability to assess the scientific merit of any work objectively and fairly.

10.8 Self-review and assessment

Hopefully you self-review on a fairly regular basis by asking questions like 'How am I doing on this module?' or 'Am I making the most of these laboratory classes?' Perhaps the most effective way you can do this is to keep a learning log or journal in which you reflect on what you have learned, review your performance and think about how it can be improved. For example, the learning log can be used to record feedback you have obtained from tutors along with how you intend to respond to the feedback, leading to improved performance in the future. Another useful, reflective approach you might adopt is to consider the advice you might give a student who is about to attempt an assignment you have just completed.

A further approach your tutor may adopt, which is essentially an extension of peer assessment procedures, is to ask you to mark your own work against the correct answers he or she will provide. As with peer assessment, this sort of participation in the marking process can be very valuable as you will receive immediate feedback and an indication of where you may have gone wrong with your answers.

10.8.1 Tutor notes

There is an extensive literature that describes learning benefits associated with self- and peer assessment, and a good introduction to this area is provided in the UK Centre for Bioscience's publication *Self- and Peer-Assessment: Guidance on Practice in the Biosciences* (Orsmond, 2004). The guide includes seven case studies that describe real-life examples of how academics have used peer or self-assessment. Despite all of this, many academics remain less than enthusiastic about adopting these procedures. Given their concerns it is important to note that the results of a number of studies confirm that students can assess their peers (for example, see Hughes, 1995; Topping, 1998) or themselves (Boud and Falchikov, 1989; Orsmond *et al.*, 1997) effectively. For a summary of good practice in the management of peer assessment see Langan and Wheater (2003).

10.9 Bringing it all together

It is all too easy to view the completion of assessments as a series of chores: tasks that will lead to a collection of marks that your university will combine, then award you a degree. Such a view will prevent you from making the most of your studies. What we hope we have shown in this chapter is that assessment is part and parcel of effective learning. We have focused in particular on the themes of formative, peer and self-assessment, and feedback. But how do these themes come together in reality? The case study below is an example of an assessment exercise based on a real assignment at the University of Hull (kindly provided by Dr Graham Scott) – you may encounter something similar in your own studies.

10.9.1 Case study

Alec worked in a group of four and contributed 2000 words to the group's report on 'The rights and wrongs of the Faroe Islands whale cull'. Before the group received a mark and feedback for the report each member had to complete an assessment exercise ('Individual Assessment 1', outlined below) worth 10% of the mark for the module.

Individual assessment 1

In the module handbook you have been provided with a copy of the assessment criteria that will be used by your examiners to grade your group report. Using these criteria as a guide:

• What grade band does your work fall into? (Tick one box.)

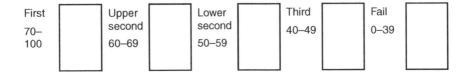

| First 70–100 | | Upper second 60–69 | | Lower second 50–59 | | Third 40–49 | | Fail 0–39 | |

• Within the band you have selected, what percentage mark would you award the work? ....%

• Justify the mark that you would award your group. In doing so you should make reference to the table of assessment criteria you have been provided with. You should also provide evidence to support your view. Remember you are assessing the submitted report – the product of your **group's** effort – **not** the work that you did as an individual. As a guide you should aim to write approx. 500 words.

In this exercise Alec had to apply the criteria he was given for the group report. He was expected to assess the whole report taking into account both his own contribution (self-assessment) and the contributions of the other group members (peer assessment).

His tutor then gave Alec and the rest of the group a mark for the report, along with feedback. His next task (worth a further 10% for the module mark) was as follows:

Individual assessment 2

(1) Individual assessment 1 required you to grade your first group report. Now that you know the grade that your tutor has awarded the work, reflect upon the similarity/difference between the tutor's mark and your mark. You should aim to write around 100 words and you might reflect on the usefulness, or otherwise, of the assessment criteria (10% of available marks).

(2) What were the three most useful pieces of written feedback that your tutor gave in relation to your first group report, and why? (approx. 50 words each; 15% of available marks).

(3) Produce an action plan to show how you will use the feedback provided by your tutors specifically to improve your second group report (approx. 250 words; 25% of available marks).

(4) Produce an action plan to show how you will use the feedback provided by your tutors in relation to other modules and to your life after graduation (approx. 250 words; 25% of available marks). (Although there may be overlap in your response to questions 3 and 4, we do have the expectation that they will differ significantly.)

(5) Justify the usefulness or otherwise of the two individual assessment components (approx. 250 words; 25% of available marks).

Through a formal assessment exercise, the tutor helped Alec make best use of the feedback he had received. The exercise did not omit discussion of the mark awarded for the group report, but tellingly only 10% of the available marks for Individual assessment 2 (IA2) were allocated for discussion of the overall mark. Instead the tutor signalled the importance of feedback by awarding most of the marks for this exercise for awareness and use of feedback. The second question in IA2 in this exercise picked up on a point made in Section 10.5.1: that sifting major and minor feedback points is a useful habit to employ. Half of the marks for IA2 were for action plans: the first related to a similar task in the same module; the second cast the net much more widely, involving other modules and life after graduation.

The final task Alec was asked to perform involved reflection on the usefulness or otherwise of Individual assessments 1 and 2. It is perhaps tempting to dismiss this task as unimportant, but arguably it is the one that best identifies an individual's self-awareness and understanding of their educational journey.

Tutors can help you up to a point, but ultimately **you** have to take responsibility for your own learning. Part of this involves you appreciating how you can benefit from the assessment process. You can then decide how well you are doing, identify areas in need of attention and finally decide how you might improve. These skills will not only equip you for academic success; they will also ensure you are employable in a wide range of workplace settings.

10.10 How you can use assessment, feedback and review to help you achieve your potential

- Before you start any assignment make sure you understand the **marking criteria** and **learning outcomes**.

- Take **formative** (as well as summative) assessment **seriously**.

- Good **feedback** is **priceless**: make the most of it.

- Your classmates have a lot to offer – participate enthusiastically in peer review/assessment.

- Critical evaluation is an essential part of being a scientist – use every opportunity to develop this skill.

- Avoid plagiarism at all costs.

10.11 References

Bellon, J.J., Bellon, E.C. and Blank, M.A. (1991) *Teaching from a Research Knowledge Base: A Development and Renewal Process* (Facsimile Edition). Upper Saddle River, NJ: Prentice Hall.

Boud, D. and Falchikov, N. (1989) Quantitative studies of student self-assessment in higher education: a critical analysis of findings. *Higher Education* **18**, 529–549.

Brown, S. and Smith, B. (1997) *Getting to Grips with Assessment* (SEDA Special No. 3). Birmingham: Staff and Educational Development Association.

Glover, C. and Brown, E. (2006) Written feedback for students: too much, too detailed or too incomprehensible to be effective? *Bioscience Education* **7**, 3. www.bioscience.heacademy.ac.uk/journal/vol7/beej-7-3.aspx.

Hughes, I.E. (1995) Peer assessment of student practical reports and its influence on learning and skill acquisition. *Capability* **1**, 39–43.

Hughes, I.E. (2001) But isn't this what you're paid for? The pros and cons of peer- and self-assessment. *Planet* **2**, 20–23.

Langan, A.M. and Wheater, C.P. (2003) Can students assess students effectively? Some insights into peer assessment. *Learning and Teaching in Action* **2**(1). www.celt.mmu.ac.uk/ltia/issue4/langanwheater.shtml.

Lunt, T. and Curran, J. (2010) 'Are you listening please?' The advantage of electronic audio feedback compared to written feedback. *Assessment and Evaluation in Higher Education* **35**(7), 759–769.

Merry, S. and Orsmond, P. (2008) Students' attitudes to and usage of academic feedback provided via audio files. *Bioscience Education* **11**, 3. www.bioscience.heacademy.ac.uk/journal/vol11/beej-11-3.aspx.

Northcliffe, A. and Middleton, A. (2008) A three year case study of using audio to blend the engineer's learning environment. *Engineering Education* **3**(2), 45–57.

Orsmond, P. (2004) *Self- and Peer- Assessment: Guidance on Practice in the Biosciences*. Leeds: The Higher Education Academy Centre for Bioscience.

Orsmond, P., Merry, S., Reiling, K. (1997) A study in self assessment: tutor and students' perceptions of performance. *Assessment & Evaluation in Higher Education* **22**(4), 357–369.

Race, P. (2005) *Making Learning Happen: A Guide for Post-Compulsory Education*. London: SAGE Publications.

Race, P. (2008) Building on Feedback. Available from http://phil-race.co.uk/students/. Accessed April 2011.

Race, P. (2009) Assessment. In UK Centre for Bioscience Briefing: Assessment, Ed. Clark, K. and Meskin, S., p.1 (www.bioscience.heacademy.ac.uk/ftp/resources/briefing/assessbrief.pdf). Accessed April 2011.

Topping, K. (1998) Peer assessment between students in colleges and universities. *Review of Educational Research* **68**, 249–276.

10.12 Additional resources (accessed April 2011)

ASSET Project. This project has a number of resources relating to the usefulness of video media for enhancing the quality and timeliness of feedback provision. www.reading.ac.uk/videofeedback/Whatisasset/.

ASKe CETL. ASKe stands for Assessment Standards Knowledge exchange, and the project has helped develop a shared understanding of assessment standards. The website contains a range of innovative resources including the popular 123 leaflets. www.brookes.ac.uk/aske/.

Assessment Audit Tool. The Tool is a good way to initiate discussion and facilitate departmental development of assessment procedures. It is designed to help teachers consider the content and design of a course/ programme with respect to a particular issue, and identify strengths and areas where improvements could be made. Available at: www.bioscience.heacademy.ac.uk/resources/audit.aspx.

Engage in Feedback website. Developed at the University of Reading, this website provides evidence-based ideas, tools and resources to enhance student feedback. www.reading.ac.uk/internal/engageinfeedback/.

Student Short Guide: *Assessment*. Advice and guidance for students on assessment in university bioscience programmes. www.bioscience.heacademy.ac.uk/ftp/resources/shortguides/assessment.pdf.

Student Short Guide: *Feedback – Make it Work for You*! Advice for students on how to make the most of feedback. www.bioscience.heacademy.ac.uk/ftp/resources/shortguides/feedback.pdf.

UK Centre for Bioscience *Briefing: Assessment*. The Briefing brings together assessment resources for bioscientists. www.bioscience.heacademy.ac.uk/resources/briefings/assessment.aspx.

UK Centre for Bioscience *Bulletin: Feedback & Feed-Forward*. A collection of articles written mainly by bioscientists, who outline how they implement good assessment and/or feedback in real-life situations. www.bioscience.heacademy.ac.uk/ftp/newsletters/bulletin22.pdf.

11 Communication in the biosciences

Joanna Verran and Maureen M. Dawson

11.1 Introduction

When you first arrive at university to study bioscience, you bring with you a range of skills that you have acquired throughout your school and college career. These include subject-specific skills acquired from all the different subjects you have studied (for example, you may be able to speak a foreign language, discuss the causes of the Second World War or draw a cross-section of a mammalian eye) and generic or transferable skills, which can be applied to situations across the subject disciplines, and in 'real life'. Generic/transferable skills include teamwork, leadership, problem solving, time management, numeracy, ICT literacy, adaptability, commercial awareness and communication. During your time at university you will be expected to improve the skills you already possess, and learn new ones.

Subject-specific skills can be acquired as a result of attending lectures, taking an active part in tutorials and undertaking laboratory practicals and projects. In addition, you will be expected to research, complete and submit assignments throughout your programme of study. The assignments you will be given will have several purposes. Though some will be used merely as a way of grading, most assignments will be designed to allow you to acquire and practise both specific and generic/transferable skills.

The ability to communicate effectively, in a range of formats and to a wide range of audiences, is one of the most important sets of skills that you can acquire as a scientist. In this chapter we will first consider why good communication skills are essential to the modern scientist. We will then introduce the range of communication skills that will be required of you during your programme. It is not our intention to provide you with a complete 'How to . . .' guide. For a start, there is insufficient space in a single chapter to do this. In addition, there are already several publications which are adequate for this purpose, and some of these are detailed within the reference section (Divan, 2009; Dawson et al., 2010). We have illustrated the chapter with several case studies to demonstrate the types of activity that can be used to generate these skills. You may find that the style of some of these exercises is already familiar if you have started your programme. However, during your course you will also find that the diversity of assignments and assessments encountered reflects the talents and interests of your academic tutors; so be prepared for something different!

Effective Learning in the Life Sciences: How Students Can Achieve Their Full Potential, First Edition.
Edited by David J. Adams.
© 2011 John Wiley & Sons, Ltd. Published 2011 by John Wiley & Sons, Ltd.

11.2   Communication skills in the undergraduate curriculum

You will require good communication skills during your programme in order to undertake, and succeed in, your assignments. At the same time, your assignments will be designed to allow you to acquire, and practise, communication and other skills. You will no doubt be familiar with conventional assignments, such as writing essays and laboratory reports, but may be surprised by the range of additional types of assignment that you will be given, some of which are described in more detail in this chapter. Table 11.1 lists a range of assignments that you may be given in order to improve and demonstrate your communication skills. We have classified communication skills as *written*, *oral* and *visual*, though there is obviously a good deal of overlap between the three forms. In addition we will discuss communication skills which revolve around *public engagement*; that is, communicating the purpose and results of scientific enquiry to the general public.

11.3   Opportunities to develop communication skills

The skills you are encouraged to acquire during your programme may not always seem to be directly relevant to your chosen subject of study. However, when you eventually begin to search for employment, these skills will contribute to the array of talents that you will bring to your employer; so it is important that you recognise and nurture them, and use them appropriately with the different audiences that you will encounter.

   You are likely to have the opportunity to volunteer for activities outside formal, timetabled contact periods, but still within your subject area. For example, you could get involved in outreach work, perhaps communicating your scientific knowledge to children in schools, or you may have the opportunity to coach students in secondary education. Alternatively, you may become a student representative, consulting with your peer group and communicating their concerns to academic staff. You may also undertake laboratory experience within a company, either as part of your degree programme or organised by yourself during the vacations. All these activities should provide opportunities for you to enhance your communication skills and your future employment prospects by building your *curriculum vitae*. Increasingly, Bioscience students are being given the opportunity to work with people from other disciplines, towards a common goal, and this gives them the experience of working in multidisciplinary teams (see www.sicb.org/careers/faqs.php3). Taking a full part in this diversity of experience will also enrich you as a working scientist. In addition, you will develop more as an individual if you improve your ability to communicate with people working in other disciplines. Make the most of all the opportunities available to you during your degree course and you will develop yourself as a unique and confident individual. Do not think that the range of communication skills you acquire will no longer be required on completion of your course. When employers are asked what they look for in a graduate, they invariably cite good communication skills as a desirable attribute (for example, see http://ww2.prospects.ac.uk/cms/ShowPage/Home_page/ What_do_graduates_do__2008/What_do_employers_want_/p!ebfpppd). This does not mean that they are not interested in your class of degree, but it is certainly true that graduates who can communicate effectively are more likely to be successful in job applications and in job interviews. It is also fair to say that students who develop good communication skills are more likely to obtain a good degree, since so much of their coursework relies on developing and demonstrating these skills.

   If you eventually choose employment as a scientist, you may be required to communicate orally, visually and in writing in different situations and to diverse audiences. Examples could include discussions with colleagues either within your workplace or with fellow scientists in collaborating establishments. You may be expected to present your work at conferences, to write reports and papers for presentation within and outside your institution. You may also be involved in training

Table 11.1 Communication skills developed and practised during assignments

Assignment type	Examples	Communication skill developed and practised			
		Written	Oral	Visual	Public engagement
Essay	Length- or time-constrained essays and exam questions	✓			
Lab reports	Report handed in at end of lab session; full report handed in several days after practical	✓			
Abstracts/summaries	Abstract/summary of journal paper	✓			
Application forms	Application for grant funding, or for real or imaginary job	✓	✓	✓	
Student journals	Students submit a report of their own project work in the format of a journal paper; students review an article submitted to a student journal (real or imaginary)	✓	✓	✓	
Information booklets	Students produce a leaflet to inform patients/families about, for example, a particular disease	✓		✓	✓
Blogs and wikis	Students discuss a scientific topic on a blog or wiki	✓	(✓)	✓	
Presentations	Individual or group presentations to peers and academic staff	✓	✓	✓	
Research conferences	Students present project work as an oral presentation at a student conference	✓	✓	✓	
Posters	Students produce a poster, either individually or as a group, on an allocated topic	✓		✓	
Student ambassadors	Students act as ambassadors at departmental open days		✓		✓
Outreach	Students coach school pupils		✓		✓
Interviewing	Students are members of interview panels, for example, in role play		✓		
Committees	Students represent their peers on university committees		✓	✓	
Presentational software such as PowerPoint	Students produce a poster, oral presentation or handouts	✓	✓	✓	
Artwork/graphics	Students produce artwork related to biological topic			✓	
Debate	Students form part of team to debate, for example, ethical issue in biology		✓		
Dragon's den	Students in groups pitch their ideas to a professional team of potential funders		✓	✓	

junior staff, or students on placement, or required to discuss your work with members of the general public, such as at open days. Some scientists are regularly called upon to talk to the media about, for example: the significance of a new discovery; an emerging disease, such as swine flu; new drugs to treat disease; or to explain a new test, for example, to detect DNA in forensic samples. Be more alert to scientists on the radio or television – note which ones communicate well, and try to pinpoint what they are doing right.

If you are not employed as a scientist, then the skills you have acquired alongside your scientific training may become even more important!

11.4   Written communication

At all stages in your degree programme you will be required to complete written assignments. The type of assignment, and the degree of difficulty, will vary according to the level at which you are studying. The most common assignments are essays and laboratory reports. However, other less conventional written assignments are used increasingly in bioscience and other degree programmes. For example, we know of at least one department in which students are given a choice of whether to submit an assignment as a written essay or as a podcast (Scown, 2008). Of course, the style of writing used for each assignment will depend on the audience to which it is directed, and this should be stated clearly in the assignment brief given to you by your tutor.

In addition to essays and laboratory reports, you may be required to produce abstracts and summaries (of your own work or the work of others), a dissertation or project report (usually in your final year) and a *curriculum vitae*. You may also be involved in writing or co-writing scientific papers, perhaps for submission to a student journal such as *Bioscience Horizons* (see http:// biohorizons.oxfordjournals.org/ and Section 11.4.3). In addition, your tutor may encourage you to submit an essay for a student prize of the type awarded by the Higher Education Academy's UK Centre for Bioscience (see www.bioscience.heacademy.ac.uk/funding/essay/award10.aspx) or to promote science communication and science literacy, for example, by working with Beacons for Public Engagement (www.publicengagement.ac.uk/beacons). Other assignments you might be given include writing press releases, information leaflets or producing and/or contributing to blogs or wikis. You will also have to complete application forms, for example, when applying for an industrial placement during your course or for a job at the end of your degree programme.

11.4.1   Essays

Despite the diversity of types of assignment used routinely by universities, the essay remains a common form of written communication in undergraduate programmes. You will be expected to write essays throughout your programme of study. In addition to coursework, most written examinations will contain at least a few essay-type questions. Usually first-year essays will be more straightforward, and of lower word limit than subsequent years. Initially, you will use textbooks as sources of information, but as your studies progress you will be expected to place much greater emphasis on primary sources of information, such as reviews and original papers published in scientific journals. The key to writing a good essay is to allow yourself sufficient time to complete it; preparation time is almost more important than the actual writing time. Make sure you know all the assignment details, including title, word limit and submission date. The assignment brief may also detail the learning outcomes of the assignment; these will be useful to you both for the assignment and for your personal reflection. A plan should be an integral part of your preparation. Take a good look at your plan: is the structure logical and is the content relevant to the title? If so, go ahead, but make sure you refer back to the plan from time to time to ensure you don't drift away from the theme and layout of the essay.

11.4.2 Blogs and wikis

A blog, (a blend of the term 'web log'), is a type of website to which an individual, or a group, can post messages. This allows a continuous online discussion to be maintained, with 'posts' appearing in reverse chronological order. A wiki is a type of website that allows the easy creation and editing of a number of interlinked web pages via a web browser. Most students are already familiar with the wiki called Wikipedia (from which the definition was obtained) when they arrive at university. You may already have used it to obtain information, without realising that you could actually edit the information that is written there. We have heard of a student assignment (not in biosciences) in which students were required to post inaccurate material to Wikipedia, and to trace the time that it took for the 'wrong' material to be spotted and edited by another subscriber. This, perhaps, explains why academic staff will tell you to be cautious when using this open encyclopaedia. You will find more detailed consideration of issues associated with blogs and Wikipedia in Chapter 8.

Blogs and wikis are used in some departments to stimulate group work and discussion.

11.4.2.1 Case study

Students in a 200-strong cohort were assigned to 'blog groups' of 7–10 in order to debate ethical case studies (Cooper and Boddington, 2005). The students had online access to 23 case studies, accompanied by lecture material and references, and each student was expected to post a critique of a single study to the group blog. This critique carried 60% of the assignment marks. In addition each student was required to post comments on the critiques posted by two peers; each of these 'critiques of critiques' carried 20% of the assignment mark.

This type of assignment allows students to develop obvious skills such as critical assessment. It also allows them to develop skills of self-critique, by comparing their own work with that of their peers.

11.4.2.2 Case study

Students on an undergraduate human biology programme were organised into groups of three and assigned a topic. Each group was required to research the topic and develop a wiki based on the results of their investigations. In addition, each group was asked to develop a poster to present, online, to the other groups (see http://bohone.wikispaces.com/). This type of assignment allows students to develop their team-working skills and to recognise the pitfalls, as well as the advantages, of group work.

11.4.3 Student journals

A number of universities, particularly in the USA, produce journals of undergraduate research. It is well worth pursuing any opportunity you may have to publish your undergraduate project work in one of these journals – see also Chapter 6. The journals vary in their remit. For example some invite contributions solely from students within the university, while others receive contributions from across the world. The advantages for a student whose work is published are manifold. Having a publication while still an undergraduate boosts both self-esteem and job prospects, and the student gains invaluable experience of writing to very strict guidelines. Often, the journals are published and reviewed both by students and staff. You might like to look into the possibility of joining the editorial team for your local undergraduate journal. This would give you valuable experience of the technical aspects of publication, and the opportunity to acquire the skills involved in peer review.

Bioscience Horizons (http://biohorizons.oxfordjournals.org/) is an online journal produced by a consortium of UK universities and supported by Oxford University Press. The journal publishes undergraduate research from students in the UK and Republic of Ireland. Students are nominated by Heads of Department on the basis of research carried out during their undergraduate projects. Abstracts are submitted, and papers are then commissioned from a shortlist by the editorial board. Students are invited to submit a paper of their research which must adhere to journal guidelines for format.

Reinvention: A Journal of Undergraduate Research is a peer-reviewed journal produced, edited and managed by students and staff at the University of Warwick and Oxford Brookes University (see http://www2.warwick.ac.uk/fac/soc/sociology/rsw/undergrad/cetl/ejournal/). The journal invites research papers from students, or student/staff collaborations, worldwide.

The Centre for Research-Informed Teaching at the University of Central Lancashire publishes *Diffusion: The UCLan Journal of Undergraduate Research*. The journal is jointly edited by an academic and a student, and publishes outstanding work by students who are nominated by their tutors.

11.4.4 *Tips for scientific writing*

For most of your written assignments, and especially for essays and laboratory reports aimed at a scientific audience (usually your tutor), you will be required to use scientific writing. A popular, journalistic style will be actively discouraged. While further guides to scientific writing are available (for example: Anderson and Poole, 1998), there are a few basic rules that you should adhere to:

(1) Always plan your work carefully before you begin writing, as time spent planning is time well spent (actually, this applies to any assignment you are given).

(2) Research the topic thoroughly before you begin. Take advantage of the facilities offered by your university library and its staff, who are there to help you locate relevant material. They will show you, often during induction week(s), how to search the library catalogue and use the available electronic sources. Make the most of their advice and be aware of the limitations of using online searches (see also Chapter 8).

(3) Always write in a clear, precise manner. Each statement should be unambiguous and there should be no 'padding'. Your style of writing should be simple, and not sensational or journalistic. Sentences should not be overlong or complicated to the extent that the meaning or emphasis is lost. For example, in the following sentence:

> *The mitochondrion, which is only found in eukaryotic cells, which differ from bacteria in having cellular contents separated into membrane-bound organelles such as the nucleus and which may be found as single cells or in multicellular organisms such as animals and plants, is involved in aerobic respiration.*

the role of the mitochondrion is lost in the discussion of eukaryotic cells. It would be better to define eukaryotic cells in one sentence, then to describe the function of the mitochondrion in the second sentence.

(4) Use correct grammar, spelling and punctuation. As well as pleasing your tutor, this will make your writing much more readable. When using material from textbooks, remember that

American spelling is often different from English. When writing in the United Kingdom, always use UK English. It is always useful to use a spell-checker when submitting a word-processed assignment. However, you should ensure that the spell-checker is set to UK English. Also, remember that words that are homophones (sound the same but mean different things) will not show up as incorrect. Examples are 'there' and 'their' or 'flour' and 'flower'.

(5) Always spell out a word or words in full when using an abbreviation for the first time. For example: deoxyribonucleic acid (DNA). Thereafter you can use the abbreviation only.

(6) Do not plagiarise the work of others. You should attribute any material, including text, laboratory results or ideas to the original source. Your university will have regulations that cover plagiarism, so make sure you read and abide by them. You will be required to reference all your sources. For a more detailed consideration of issues associated with plagiarism see Chapter 9.

(7) Be prepared to keep revising drafts to improve the work and reinforce the material in your own mind.

(8) Your assignments will usually have a word limit, so make sure that you abide by it. Learning how to work within limits, whether in words, time or cash, is a useful skill to acquire. Many departments penalise overlong assignments formally, but a piece of work that contains far fewer words than the maximum number permitted is unlikely to have addressed the assignment in sufficient depth.

11.5 Visual communication

You will probably be asked to use PowerPoint technology, both to communicate visually during oral presentations and when you prepare posters, throughout your degree course. PowerPoint is easy to use and can be very effective. Other electronic visual presentation tools, such as Prezi, can be used during oral presentations but are not usually available on the 'standard platforms' provided by universities. Prezi has the advantage of being fairly novel in not using slides; it comprises a process which zooms into a region of a seemingly infinite display area that addresses a particular topic (see http://prezi.com/ for some examples).

Other means of visual communication with which you might be involved include: design of leaflets and other publications in print; the production of films; and representational or abstract art to illustrate biological concepts.

Using visual means for communication can have advantages over the strictly written approach. For example a single poster or PowerPoint slide can be viewed by a larger audience, at any one time, than can a written review or journal article. This allows for more discussion amongst members of the audience. A poster can provide a very effective means of summarising and communicating information in an easily 'digestible' form. Knowing how to summarise and display information effectively and accurately, in a manner that grabs attention, is a particularly useful skill to acquire. Remember that posters should be attractive, well organised, readable from a distance, and succinct. You will need to consider your audience, the overall message you want to convey and the size and orientation of the poster. Similarly, you will need to think carefully about the layout and appearance of PowerPoint slides used as part of an oral presentation. It is tempting to over-do a PowerPoint presentation with too many visual effects so that the message gets drowned out by the medium. It is also worth re-emphasising here that you should always quote sources of material, including diagrams, used in your presentation to avoid accusations of plagiarism.

11.5.1   Posters

Posters should be designed so that their message is clearly conveyed. For example, you may be required to produce a poster which summarises your final-year project. This would need a completely different approach, say, to a poster produced to provide information on the structure and function of a particular macromolecule, or one which is designed to provide information to the public, perhaps on a clinical condition.

Most posters that you see on billboards or walls convey particular messages succinctly. For example, the poster may contain a public health message or may be advertising a product or a concert. They need to be eye-catching and thus very well designed, and the message needs to be conveyed clearly in a manner that can be rapidly understood by the audience.

11.5.1.1   Case study

A good example of a student-focused poster exercise was one in which students were required to design posters providing information for travellers about immunisation against different infectious diseases (Figure 11.1). In this assignment students worked in groups, each group being assigned a 'disease' identified by the local Travel Clinic where members of the public sought advice about vaccination when travelling abroad. The posters were displayed at the clinic, where patients were asked to vote for the poster they liked best. The posters were also peer- and tutor-assessed during a class presentation session. A prize, in the form of a gift voucher for the winning team, was donated by a local pharmaceutical company (Verran, 1993).

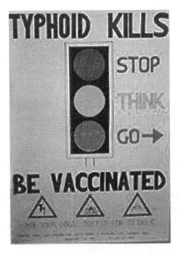

Figure 11.1 Poster conveying simple message regarding typhoid immunisation

Posters are a useful means of summarising and displaying information on one page, and key points should be easily identified, facilitated by good design (Figure 11.2). Assigning topic themes (enzymes, diseases, metabolic pathways etc.) to a class of students gives the potential for display, comparison and learning from one another in a social environment.

Posters are often used as a means of conveying preliminary research findings. Research posters typically comprise a title, brief introduction and statement of aims, methods, results and brief discussion/conclusion (Divan, 2009; Dawson *et al.*, 2010). Scientific conferences rely heavily on

RABIES — *Mononegavirales / Lyssavirus*

What is it?

Rabies is a zoonose that can be passed to humans via a bite wound (infected saliva) from a rabid animal. Dogs, rodents, skunks and bats can all be infected and are potential reservoirs for the disease.

What is the causative agent?

The disease is caused by a virus belonging to the order *Mononegavirales*, viruses with a nonsegmented negative stranded RNA genome. The virus is of the *Rhabdoviridae* family, of the genus *Lyssavirus*.
The rabies genome encodes 5 proteins:

- ❖ Nucleoprotein (N)
- ❖ Phosphoprotein (P)
- ❖ Matrixprotein (M)
- ❖ Glycoprotein (G)
- ❖ Polymerase (L)

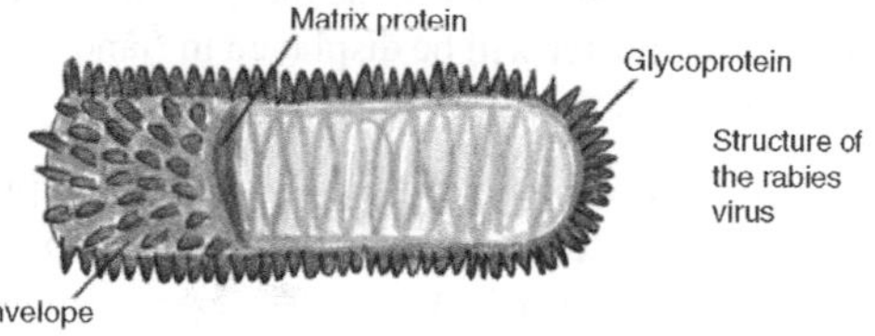

Matrix protein

Glycoprotein

Structure of the rabies virus

Envelope

Infection in humans

After the animal bite has punctured the recipient's skin, saliva infected with the rabies virus is deposited at the site of the wound. Replication occurs and the virus undergoes uptake into the peripheral nerves. The virus is then transported to the Central Nervous System via retrograde axoplasmic flow.

Symptoms of the disease

The rabies virus causes an acute encephalitis in all warm blooded hosts and the outcome is almost always fatal.
The incubation period can vary from a few days to several years, but is typically 1-3 months. After this, active cerebral infection can result in cerebral dysfunction, anxiety, confusion, agitation, delirium, hallucination and insomnia (this can vary depending on the strain of the infective virus).

Vaccination and treatment

Animals: domestic pets are routinely vaccinated against the disease, which in turn serves to protect humans from contact with the virus.

Humans: a rabies vaccine regimen can be administered to persons in high risk groups in order to provide immunity to rabies (preexposure prophylaxis).
After an exposure to rabies, postexposure prophylaxis can be administered.
After symptoms of the disease appear however, there is no treatment available.

How common is the disease in humans?

Rabies is responsible for an estimated 40-70 000 deaths world-wide per year, mainly in Eastern or undeveloped countries. The disease is endemic in many countries but not the UK.

NEWSFLASH!!!

In the UK the first case for 100 years was discovered this month, where 56 year old David McRae from Scotland was infected by the European Bat Lyssavirus (EBL), via a bite from a rabid daubenton bat. He died earlier this week.

Figure 11.2 'Essay on a poster', describing properties of rabies virus

poster presentations to supplement oral presentations, enabling many more researchers to present their work to peers. The organisers of these conferences usually issue specifications for posters including elements of the layout and overall size (although an indication of whether the poster should have portrait or landscape orientation is not always provided). Interested delegates will read the abstract for a poster in the conference proceedings (published prior to the event) and may then choose to talk to the presenter during period(s) set aside for face-to-face discussions by the poster.

At some conferences, poster presenters are asked to give a brief (two to five minute) oral explanation of the work described in the poster.

Your final-year project provides a perfect opportunity for you to present your work in the form of a research poster (Figure 11.3). Many universities formally assess posters that summarise research project studies. During the assessment you may be asked to stand by your poster and give a presentation, during which you will be expected to talk about the work described in the poster, then answer questions asked by examiner(s). Some universities have organised research conferences based around final-year project posters. It is also likely that postgraduate, postdoctoral and other researchers at your university will run internal research conferences with posters describing cutting-edge research underway in their laboratories. You should take every opportunity to attend these meetings and learn from the range of posters on display. Identify the ones you like best and decide why they are effective.

11.5.1.2 Tips for producing posters*

(1) Make sure you know the dimensions of the poster and adhere to these: this is the equivalent of a word-limit for a written assignment. The most common sizes of poster are shown in Table 11.2. Note also whether the poster will be displayed in 'landscape', with the longest sides horizontal, or 'portrait', with the shortest sides horizontal.

(2) Make sure that you know the type of poster you should prepare (for example, an 'information' poster or a 'research' poster) and who will look at it (for example, patients in a travel clinic, or staff and colleagues at university).

(3) The poster should be succinct and well organised. Do not try to cram in too much information as people will avoid an overcrowded poster. Make sure the order is logical: for example, with a research poster the results should come before the discussion section.

(4) Make sure the font size is sufficiently large so that the poster is readable from a distance of one to two metres. There is sometimes a temptation to reduce the font size so as to fit in more detail; you should avoid this temptation.

(5) Use short sentences, and paragraphs of no more than 20 lines; use lists where this is appropriate.

(6) Mix text and graphics but make sure the graphics are relevant and not there simply to provide colour.

(7) Use no more than four colours at the very most.

(8) Make sure your poster has a title and author(s).

Table 11.2 Common poster sizes

Size	Dimensions (mm × mm)
A0	841 × 1189
A1	594 × 841
A2	420 × 594
A3	210 × 297

---

* Adapted from Dawson *et al.* (2010).

In vitro studies of laboratory model biofilm susceptibility to commercial denture cleansers

Manchester
Metropolitan
University

Sutula, J. and Verran, J.
The Department of Biological Sciences, Manchester Metropolitan University,
Chester Street, Manchester, M1 5GD, UK

Introduction

Appropriate plaque control is essential for maintenance of oral soft tissue health for dentate, endentate and partially dentate individuals. It is well established that denture cleansers help to control denture-induced mucosal inflammation and stomatitis by reducing the amount of plaque.

In order to evaluate the effectiveness of denture cleansers *in vitro*, the Constant Depth Film Fermenter (CDFF) was used to produce microbial biofilms. This laboratory model closely mimics the complex microflora, the nutrient source and the substratum present in the oral cavity.

The aim of this project was to compare the antimicrobial activity of commercially available denture cleansers on the viability of mixed species oral biofilms grown using the CDFF. The susceptibility of yeast, mainly *Candida albicans*, lactobacilli and anaerobic species was of particular interest.

Materials and methods

- Multispecies oral biofilms were grown on rapid cure polymethyl methacrylate (PMMA) denture acrylic discs

- CDFF (fig. 1) described by Wilson (1999) consisted of stainless steel turntable containing polytetrafluoroethylene (PTFE) pans, rotated under PTFE scraper bars, which smeared the inoculum medium over the pans. Each pan had five cylindrical holes holding the PTFE plugs and the acrylic discs for a biofilm growth.

Figure 1. The Constant Depth Film Fermenter
(University of Wales, Cardiff, UK)

- Pooled human saliva collected from 10 volunteers was used to provide oral microbial species for the CDFF inoculum

- General and selective media as well as simple identification tests were used to explore the microbial diversity of the pooled saliva

- The inoculum was additionally spiked with *Prevotella intermedia* BH20/30, black pigmented colonies isolated from the pooled saliva and *Candida albicans* GDH 2346

- The mucin-based artificial saliva, enriched with 40% urea and 1% heat inactivated horse serum (Pratten, 1998), was a nutrient source for a biofilm development

- Fourteen-day biofilms were subjected to treatment with distilled sterile water (control) or one of the commercial denture cleansers for a duration recommended by the manufacturer:
- Steradent® 3 Minutes Fresh Mint
- Corega® Tabs®
- Polident® Fresh Cleanse
- Steradent® 10 Minutes Active Plus
- Dentural® (BDH)

Results

Pooled saliva investigation

- 98% of the total cultivable flora constituted of anaerobic species, mainly facultative Gram negative rods (fig. 2b) and coccobacilli (fig. 2a)

- Black-pigmented species comprised 0.57% of the total anaerobic population

- Lactobacillus (fig. 2f) comprised around 0.5%, and yeast including *Candida albicans* (fig. 2d,e) composed an even smaller fraction of the percentage total viable count

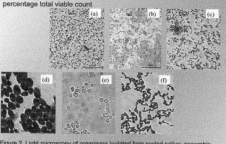

Figure 2. Light microscopy of organisms isolated from pooled saliva: anaerobic species: a) Gram negative coccobacilli, b) Gram negative rod, c) Gram positive diplococci; yeast: d) Gram stained yeast cells, e) *C. albicans* producing true hyphae; lactobacilli: f) Gram positive long rod

Denture cleansers testing

- Denture cleansers reduced microbial viability compared to water (control)

- All tested products had similar antimicrobial properties

- There was a significant difference in reduction of anaerobic species between tested cleansers and control (P< 0.05)

- Denture products reduced viability of lactobacilli and *C. albicans* although no significant difference was observed

Discussion

The work carried out in this project confirmed antibacterial and antifungal properties of commercially available denture cleansers on mixed species biofilms developed using the CDFF. All denture cleansers significantly reduced microbial viability within produced biofilms. The statistical analysis did not confirm a significant decrease in the viability of yeast including *C. albicans*; however the cleaning significantly reduced the number of anaerobic species. The cultivable microflora of fresh pooled saliva was dominated by facultative anaerobic species, predominantly Gram positive rods and cocci, with only small proportions of lactobacilli and yeast, including *C. albicans*. These results would be in agreement with cultural studies on denture plaque in patients with healthy oral mucosa (Theilade *et al.*, 1983).

References

Willson, M. (1999) Use of Constant Depth Film Fermentor in studies of biofilms of oral bacteria. *Methods in Enzymology*, **310**: 264-279.
Pratten, J., Willis, K., Barnett, P. and Wilson, M. (1998) *In vitro* studies of the effect of antiseptic-containing mouthwashes on the formation and viability of *Streptococcus sanguis* biofilms. *Journal of Applied Microbiology*, **84**: 1149-1155.
Theilade, E., Budtz-Jorgensen, E. and Theilade, J. (1983) Predominant cultivable microflora of plaque on removable dentures in patients with healthy oral mucosa. *Archives of Oral Biology*, **8**: 675-680.

Figure 11.3   Undergraduate project poster, in research conference format

(9) Include a reference list (to avoid accusations of plagiarism) and a list of acknowledgements (for example, if you have had help from a clinician when researching a disease, or a technician when undertaking your project).

11.5.2 Art and the communication of science

Some scientists, probably more than you imagine, are also interested in the arts. You might think that scientists and artists are unlikely to work together but, in fact, scientists and artists frequently join forces in 'sci-art' collaborations to convey principles and ideas through art and the arts. The Wellcome Trust sponsors a number of collaborations of this nature (see www.wellcome.ac.uk). Many of you will have come to university with interests and expertise in art, photography, design, fashion, music and so on. You may well have an opportunity to develop these skills during your studies in the biosciences.

11.5.2.1 Case study

At Manchester Metropolitan University, first-year bioscience students attend a lecture outlining examples of the links between microbiology and art that range from biodeterioration of cultural heritage, through the beauty of many biological images (look in any textbook for examples), to more overt sci-art collaborations. Afterwards, the students can choose to undertake an assignment on the topic, the outcome being a product that links microbiology and the arts. The creative outputs are varied and impressive (Verran, 2010), and have been used in internal and external publicity material, as education resources in schools, for decorative purposes in university buildings and as inspiration for subsequent projects (Figure 11.4).

Figure 11.4 (a) Graphic art depicting bacterial motility. The work was accompanied by a portfolio that displayed the research the student had collated, and explained the process used to create the artwork (Georgina Phelan). (b) Fashion designs that represent some of the properties of malaria. The student produced a portfolio, including background information on the disease, alongside sketches and fabrics for dresses 'inspired by malaria'. One finished garment was submitted alongside the portfolio (Marya Akhtar). (c) Skin biofilm cake, accompanied by recipes (Jessica Murray)

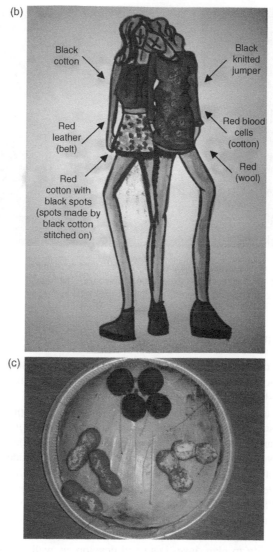

Figure 11.4 *(Continued)*

Similar projects have incorporated design considerations into the production of leaflets in line with briefings from external 'clients' (Verran 1992). Students also participate in interdisciplinary projects, where art and science undergraduates work together to produce a piece for display in the Faculty of Science and Engineering foyer (Dawson *et al.*, 2005; Figure 11.5). On other occasions, students have organised their own photography competitions. So remember that the skills you bring to university, which do not seem to be directly related to your degree subject, may prove to be invaluable during your studies and may ultimately be a very important part of your portfolio!

Figure 11.5 A mock-up of a sculpture produced by art and science students for display in the Faculty of Science and Engineering at Manchester Metropolitan University. The repeating, yet changing structure represents phenomena encountered in different disciplines (fractals, evolution, mutation, polymers etc.)

11.6 Oral communication

You will be expected to communicate orally with your peers and teachers throughout your university career. This will include the informal discussions that you have with members of your peer group, perhaps during group-based activities, and the discussions that take place within interactive tutorials. You may be asked to summarise a piece of work in a brief presentation given during such tutorials. Oral presentations are often assessed and you may be asked to give these presentations as an individual or as part of a team. The audience may be the other members of your seminar group or the entire cohort for your programme and year; for example during a student research conference. You will be given a strict time limit, and will be graded on a number of qualities, including adherence to the time limit, the relevance of the subject matter and the manner of your presentation. The latter will include style (see Section 11.6.3, Tips for oral communication, below), audibility and your ability to answer any questions. Many students find formal oral presentations nerve-wracking, and may try to avoid them. Though scary, such presentations develop skills useful for future employment, including the ability to plan, investigate and develop, as well as communicate, a theme. They are excellent for developing confidence.

As with all of your assignments, it is essential that you plan well in advance so that you are confident about the material you will use. You should also rehearse by giving the presentation in an empty room or, better still, to your flatmates. Ensure that you adhere to the time limit. A presentation that is too short is as bad as one that exceeds the time allotted. Make sure that you are familiar with the correct pronunciation of technical terms, and prepare yourself for any awkward questions that might arise.

In the modern university, with large class sizes and modular courses, there are, unfortunately, few opportunities for students to communicate orally with individual members of staff and develop one-to-one relationships. Most universities operate a personal tutor scheme, particularly in the early years of a degree programme. However, it is likely you will be expected to wait until the final year of your studies before you get to know an academic member of staff, usually your project supervisor, fairly well. This should not stop you taking the initiative and if, for example, there is a lecturer whose work is of particular interest to you, you should make the effort to talk to him or her! Most academic staff will be really happy to help, and get to know, individual students, and may be able to offer you advice, work experience or appropriate contacts that will help develop your particular interests. This mature interaction is a two-way process, and it is important that you recognise and appreciate the time and effort your tutor or other member of staff may devote to your personal development.

Talking to tutors and peers is an essential part of student life, but there will be additional opportunities to improve your oral communication skills. For example, you could: organise a study group; be a mentor for students in earlier years; ask a research team leader if you can join their group and attend their seminars; join or set-up a student society; become a STEM ambassador (www. STEMnet.org.uk) and act as a role model for young people interested in science, technology, engineering and maths; volunteer to help at University Open Days as a student ambassador. For all of these activities you will be expected to talk about your subject of study to different audiences. More formally, there are organisations that can arrange school teaching experience (www.tda.gov.uk). Finally, you could look into whether your university offers student employment opportunities or organises volunteering activities in the local community or elsewhere.

11.6.1 Podcasting

A podcast was first defined by the Oxford English Dictionary in 2005 as 'a digital recording of a radio broadcast or similar programme, made available on the Internet for downloading to a personal audio player'. Nowadays students are very familiar with podcasts; for example through downloading music from the web or listening to programmes on BBC iPlayer. The definition has been expanded to include the use of video podcasts (sometimes shortened to vidcasts or vodcasts) which use videoclips designed to be viewed using a personal computer or mobile device.

Many university departments now prepare podcasts of learning materials, perhaps even of whole lectures, and make them available on the university's Virtual Learning Environment (VLE; see Chapter 8) for students to download. Others make 'revision podcasts' available nearer to exam time, and these are generally very popular with students (Scott, 2008).

11.6.1.1 Case study

Students are sometimes asked to produce a podcast or vidcast as an assignment. Cane and Cashmore (2008) developed a podcasting activity for second-year medical students taking a module that focused on developments in genetics and their ethical implications. Small groups of around six students were required to research a relevant topic and to produce a podcast. Examples of topics included *Designer babies* and *Genetic fingerprinting*. The podcasts, each 5–10 minutes long, were made available on the VLE for students who had not taken the module. Students who had taken the module valued the experience of communicating material in this novel format.

11.6.2 Student conferences

Many departments require that students present the findings of their final-year project at a formal, undergraduate research conference. The style is that of a full conference, with students producing abstracts, presenting their paper orally in an allocated, strictly time-limited slot, and answering questions posed by a mixed audience of peers, postgraduate students and academic staff. The experience is enhanced if the conference is held away from the university, at a proper conference centre, such as Gregynog Hall used by the University of Wales. It may be very daunting, but most students enjoy (eventually) the opportunity to present the results of their project in this way. Some departments may hold a mixed undergraduate/postgraduate conference at which both undergraduates and postgraduates present their research findings; this gives undergraduates the opportunity to talk to postgraduates about their work.

11.6.3   Tips for oral communication

(1) Make sure you are clear about the purpose and length of the presentation, and the target audience.

(2) Research the topic thoroughly and allow sufficient time for preparation of the presentation.

(3) Make sure the presentation is ordered logically into introduction, main body of information, and concluding remarks or summary.

(4) Check the facilities available; for example, if you need a projector with computer make sure you have access to these.

(5) Practise the presentation by giving it several times to an empty room or to friends; ask them to listen and critique the content and delivery.

(6) Make sure you maintain eye contact with your audience: never turn your back on them to deliver the presentation to the projector screen (a frequent mistake made when people are nervous).

(7) Make sure your audience can hear you by speaking clearly and sufficiently loudly (in a large room a good tip is simply to ask the people at the back if you can be heard).

(8) Deliver at a pace that is not too fast and not too slow (if you have rehearsed with an audience of friends then they should have advised you on this). There is a temptation when nervous to speak too rapidly – if you have rehearsed well you should be able to overcome your nerves.

(9) Never read out your presentation to your audience: this leads to a monotonous tone and lack of eye contact.

(10) Make sure you know how to pronounce scientific terms before the presentation – this will add to your confidence.

(11) When using PowerPoint slides, take advice from your tutor on font size and display options – do not be tempted to wow your audience with your technical abilities at the expense of the actual content. Do not display too much information and never read from the slides – use them as a prompt only – the slides should complement the content of your talk.

(12) At the end of the presentation, thank your audience and ask if they have any questions: thorough preparation should mean you are able to answer them.

11.7   Public engagement

In recent years, universities have been expected to be more accountable to the public. The Beacons for Public Engagement project (www.publicengagement.ac.uk/beacons) was set up to encourage universities and their immediate neighbourhoods to interact more actively. Initiatives of this nature clearly provide students with opportunities to exercise and develop their communication skills. Similarly you might wish to participate in local science festivals or events run by museums; the organisers are likely to welcome student input into their public engagement events. Associated activities such as writing press releases, giving radio interviews, and developing engaging and educational tools will also enrich your communication skills portfolio. Social media were defined by Kaplan and Haenlein (2010) as a 'group of Internet-based applications that... allows the creation and exchange of user-generated content.' They include blogs, wikis and sites such as Twitter (http://twitter.com) and Facebook (www.facebook.com), and provide additional and novel opportunities for demonstrating your awareness of new developments whilst widening public engagement with science. For example, you could set up, advertise, contribute to and monitor a Facebook site to discuss scientific topics.

Always be guided by your tutor when engaging with the public as a representative of your university. You should take advantage of any training that is offered by your institution. Indeed, the university may insist that you do so.

11.7.1 *Case study*

The Bad Bugs Book Club (www.sci-eng.mmu.ac.uk/intheloop) was set up to encourage discussion about novels in which infectious disease formed part of the plot; the aim being to encourage reading, to extract and evaluate scientific principles, and to enhance public understanding of, and interest in, science. A range of genres has been explored including historical and science fiction; and films of the book in question, where available, have been viewed and compared. For each book, a reading guide has been produced as well as a report of the meeting at which the book is discussed. Book club membership encompasses scientists and non-scientists, facilitating exploration of the book's scientific content in an informal environment. Students have been members of the book club, and there is obvious potential for incorporating a book club activity in an undergraduate module. In addition, a range of scientific subjects can be the focus of the book club which, of course, does not need to be confined to books involving microbiology. This activity encourages wider reading, interdisciplinary links and the development of oral communication skills in a relaxed setting.

11.7.2 *Case study*

The Ideas Foundation (www.ideasfoundation.org.uk) works with design agencies to encourage design skills in schoolchildren. One such project required students at an arts and science specialist college to re-design their school planner so as to reflect the cross-disciplinary expertise available within the college. The project involved a two-week 'work placement' with students from the Biology and Design departments of a University. They mentored the college students and helped them to think across disciplines, to identify messages about health that would go in the planner, and consider and implement elements of design. Science-based activities were built around a lecture on microbiology and art, during which a number of undergraduate science students described their creative outputs, and also on guided discussions around the links between vampirism and infectious disease. These rather obscure cross-disciplinary connections helped the students (both university and college) to think more adventurously in the design of the planner. The exercise culminated with three groups of college students, each mentored by a science and an art undergraduate, pitching their ideas to their College Head and a group of designers, and the best pitch was chosen for the planner. Mentoring activities such as these help undergraduates build confidence and give them the opportunity to develop a range of communication skills; in all they are invaluable experiences that can greatly enhance a student's *curriculum vitae*.

11.7.3 Case study

Sahrish wanted to do her final-year project/dissertation in an area that would help her apply subsequently for a teaching course. The aim of her project was to help organise a public engagement event to raise awareness of World AIDS Day (1st December). Additional funding from the Society for General Microbiology enabled collaborative work with the embroidery department of her university (Manchester Metropolitan; MMU), and a community quilt was produced and completed at the university on World AIDS Day. Sahrish organised the event, and produced posters providing information about HIV, the AIDS Memorial Quilt (see www.aidsquilt.org), the history of quilting and the MMU community quilt (Figure 11.6).

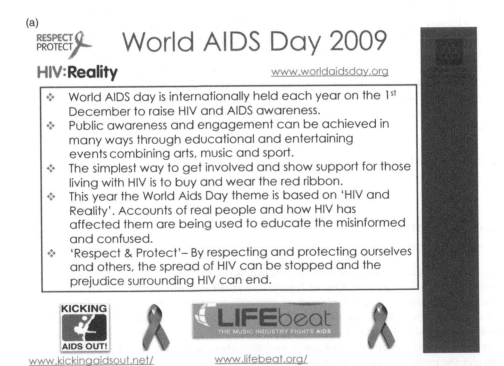

Figure 11.6   (a) Posters designed and produced to provide a backdrop for a public engagement event raising awareness of HIV AIDS. (b) Community embroidery event on World AIDS Day (sponsored by Society for General Microbiology)

HIV:Reality

What is AIDS?

❖ Acquired Immune Deficiency Syndrome or AIDS is a condition whereby the body's immune response becomes suppressed. Opportunistic pathogens are then able to take advantage of the weakened immune system to bring about a variety of infections.

What is HIV?

❖ AIDS is caused by Human Immunodeficiency Virus or HIV, which is a retrovirus that replicates within the body, destroying a subgroup of lymphocytes and ultimately the body's defence system.

❖ HIV can be contracted through semen, vaginal fluids, breast milk or infected needles and blood.

❖ People who are infected with HIV often do not show any symptoms and appear healthy. The only certain way to diagnose it is via a blood test.

❖ Antiretroviral drugs can control the amount of virus in the body, allowing the immune system to remain effective. However there still is no cure.

HIV:Reality

The AIDS Quilt

www.aidsquilt.org

❖ The AIDS quilt was founded in 1987 and is the longest ongoing community arts project in the world.
❖ 30 countries including England contribute to the panels that make up the AIDS quilt.
❖ Each panel is 90 x 180 cm and is designed as a personal tribute to a loved one lost to AIDS. It can be assembled using paint, applique, stencil and photos.
❖ Portions of the quilt are constantly on display around the world.
❖ **The aims of the AIDS Memorial Quilt:**
-A means for remembrance and healing
-To increase awareness of HIV and AIDS
-To assist others in providing education on the prevention of HIV infection
-To raise funds

Figure 11.6   (*Continued*)

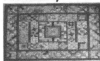

The MMU AIDS Banner

At MMU, microbiology is combining with embroidery to host a public awareness raising and remembrance event for World AIDS Day 2009.

Quilting

Quilting is the process of securing some form of padding between two layers of fabric using stitches that create a decorative design.

The communal MMU banner has been designed using cross stitching and the internationally recognised AIDS symbol.

Quiltmaking has a history of communal engagement; for example American colonial housewives, abolitionists.

The concept for the MMU AIDS banner is based on the AIDS memorial quilt, which was established over 20 years ago. It is successful in demonstrating the impact of the AIDS pandemic whilst increasing awareness, and inspiring action in the struggle against HIV and AIDS.

Different community groups have contributed their artwork and skills towards this project in creating the message 'respect and protect' which will be conveyed to the public. The banner will be completed on World AIDS Day, where members of the public are invited to participate in the final quilting session.

(b)

Figure 11.6 (*Continued*)

Sahrish also ran a book club using a book of short stories 'AIDS Sutra' as a focus for discussion amongst her peers. During this time she became a STEM ambassador, worked as a volunteer and as a 'science busker' at the Manchester Science Festival, and helped with other public engagement events at the University. Clearly the mark she obtained for her dissertation was only part of the contribution the activity made to her employability.

11.8   How you can achieve your potential as a communicator

- Use the resources and opportunities available to you to practise and improve your communication skills.

- Identify your strengths and weaknesses: use your strengths to circumvent weaknesses.

- Reflect on your experiences and adapt your activities to maximise impact.

- Recognise the different ways in which science can be communicated to different audiences and plan an appropriate communication strategy for different circumstances.

- Be aware of the importance of good communication skills in science and for your future employment.

11.9   References

Anderson, J. and Poole, M. (1998) *Assignment and Thesis Writing* (3rd edition). John Wiley & Sons Australia Ltd.

Cooper, C.D. and Boddington, L. (2005) Assessment by Blog: Ethical Case Studies Assessment for an Undergraduate Business Management Class (http://incsub.org/blogtalk/?page_id=62). Accessed April 2011.

Cane, C. and Cashmore, A. (2008) Student-Produced Podcasts as Learning Tools (www.heacademy.ac.uk/resources/detail/events/annualconference/2008/Ann_conf_2008_Chris_Cane). Accessed August 2010.

Dawson M.M., Bailey, S., Dunbar, T. and Verran, J. (2005). Use of a Sci-art project to explore the benefits of interdisciplinary collaboration. In *Proceedings of the Science Learning and Teaching Conference 2005, 27–28 June 2005, University of Warwick*, Ed. Goodhew, P, Murphy, M., Doyle, D. *et al.* HEA Subject Centres for Bioscience, Materials and Physical Sciences, pp. 62–65. Available at: www.bioscience.heacademy.ac.uk/ftp/events/sci_lt05/SLTCproceedings_05.pdf. Accessed January 2011.

Dawson, M., Dawson, B.A. and Overfield, J.A. (2010) *Communication Skills for Biosciences* Chichester: John Wiley & Sons, Ltd.

Divan, A. (2009) *Communication Skills for Biosciences: A Graduate Guide*. Oxford: Oxford University Press.

Kaplan, A.M. and Haenlein, M. (2010) Users of the world, unite! The challenges and opportunities of Social Media. *Business Horizons* **53**(1), 59–68.

Scott, N. (2008) An evaluation of enhanced student support (including podcasting) on assessed course work achievement and student satisfaction. *Learning and Teaching in Action* **7**(1). www.celt.mmu.ac.uk/ltia/issue15/scott.php.

Scown, P. (2008) Using students' assignments to create a library of re-usable learning objects. *Learning and Teaching in Action* **7**(1). www.celt.mmu.ac.uk/ltia/issue15/scown.php.

Verran, J. (1992) A student centred learning project: the production of leaflets for 'live' clients. *Journal of Biological Education* **26**, 135–138.

Verran, J. (1993) Poster design: by microbiology students. *Journal of Biological Education* **27**, 291–294.

Verran, J. (2010) Encouraging creativity and employability skills in undergraduate microbiologists. *Trends in Microbiology* **18**, 56–58.

11.10   Additional resources (accessed January 2011)

Carpi, A., Eggar, A.E. and Kuldell, N.H. (2008) Scientific Communication: Understanding Scientific Journals and Articles (www.visionlearning.com/library/module_viewer.php?mid=158&l=&c3).

Cornell Centre for Materials Research (2006) Scientific Poster Design (www.cns.cornell.edu/documents/ScientificPosters.pdf).

Monash University (2007) The Science Essay (www.monash.edu.au/lls/llonline/writing/science/6.xml).

12 Bioenterprise

Lee J. Beniston, David J. Adams and Carol Wakeford

12.1 Introduction

12.1.1 Why should a bioscience student be interested in enterprise?

Bioscience is big business! The modern biotechnology industry is relatively new and has grown to a remarkable extent during the last three decades. This growth has been based on a number of key discoveries and other developments summarised in Box 12.1. At the core of the industry are products and services commercialised from bioscience research in three main areas: healthcare, fuel and agriculture (including food). Publicly traded biotech companies in the USA in 2008 had an estimated value of $360 billion, whilst revenues from healthcare biotech in the USA increased by $50.8 billion in 14 years (BIO, 2008). Biotechnology now plays a major part in many global economies and is likely to continue to increase in economic importance in the next few decades. For example, many believe that the data generated from human genome sequencing will further revolutionise the bioscience industry. The Biotechnology and Biological Sciences Research Council suggest: 'The 21st century will be the age of bioscience. A biological revolution is unfolding' (BBSRC, 2010).

As a student of the biosciences you will have the opportunity to participate in these exciting developments. Many students have great creative potential (see Chapter 1) and it is possible that, during the course of your career, you will have valuable ideas that you will wish to exploit commercially. If you are to do this, you will have to be both enterprising and entrepreneurial, and in this chapter we provide an overview of 'bioenterprise', the process that involves the commercial exploitation of ideas and results by bioscientists. The overview is presented in three phases that will help you identify and protect an idea or opportunity (Phase 1), research the market potential of the idea (Phase 2), then translate it into a bio-business proposition by means of a business plan (Phase 3) (Figure 12.1). The skills you acquire will enable you to make the most of your own ideas and should enhance your employability in a wide range of career settings.

12.1.2 Bioenterprise and knowledge transfer

Knowledge transfer (KT; Booth and Armour, 2008) is a term you will hear a great deal in the context of bioenterprise, and it is essential you are fully aware of what is involved in this process. A common

Effective Learning in the Life Sciences: How Students Can Achieve Their Full Potential, First Edition.
Edited by David J. Adams.
© 2011 John Wiley & Sons, Ltd. Published 2011 by John Wiley & Sons, Ltd.

Box 12.1 Timeline for important discoveries and events in biotechnology

1977   A gene encoding a human protein (somatostatin) is expressed in a bacterium by
          Genentech.
1978   US scientists show it is possible to introduce specific mutations at specific sites in a
          DNA molecule.
1980   US Supreme Court approves the principle of patenting organisms. US patent for gene
          cloning is awarded to American biochemists. First gene synthesising machines are
          developed.
1981   First transgenic animals are produced by transferring genes from other animals into
          mice.
1982   Genentech receives approval from US Food and Drug Administration (FDA) to
          market genetically engineered human insulin.
1983   Polymerase chain reaction invented. First artificial chromosome synthesised.
1984   DNA fingerprinting technique is developed.
1985   Transgenic plants resistant to insects, viruses and bacteria are field-tested for the first
          time.
1986   First recombinant vaccine (for hepatitis B) approved for humans. First field tests of
          transgenic plants (tobacco) are conducted.
1988   First US patent awarded for a genetically altered animal – a transgenic mouse. Juries in
          the USA and UK deliver the first murder convictions based on DNA evidence.
1990   Human genome project formally launched. First products of recombinant DNA
          technology enter the food supply in UK and USA. First experimental gene therapy.
          First transgenic dairy cow.
1994   US FDA approves the first whole food produced through biotechnology: FLAVR
          SAVR™ tomato.
1995   First full gene sequence for living organism, the bacterium *Haemophilus influenzae*.
1996   Farmers plant transgenic staple crops, including corn, soya beans and cotton, for the
          first time.
1997   Dolly the sheep cloned.
1998   First complete animal genome, that of the worm *Caenorhabditis elegans*, sequenced.
          Technique for culturing embryonic stem cells is developed. First gene chip for
          transcriptional profiling of industrial organism is designed.
2000   Draft of human genome sequence announced. First complete map of plant genome
          (*Arabidopsis thaliana*).
2001   Single gene from *A. thaliana* inserted into tomato plants to create the first crop able to
          grow in salty water and soil.
2002   Rice is first major crop to have its genome sequenced. Synthetic virus (polio) is
          assembled using genome sequence information.
2003   China grants world's first regulatory approval of a gene therapy product.
2004   United Nations Food and Agriculture Organization endorses GM crops. First com-
          mercial production of bioethanol using recombinant enzymes and wheat straw.
2005   World Health Organization concludes GM crops can enhance human health and
          economic development (although stressing need for continued safety assessments).
          Corn-derived ethanol production at 4 billion gallons per year.
2006   First regulatory approval for plant-made vaccine (for protection of poultry).
2007   Successful reprogramming of human skin cells to create cells indistinguishable from
          embryonic stem cells.

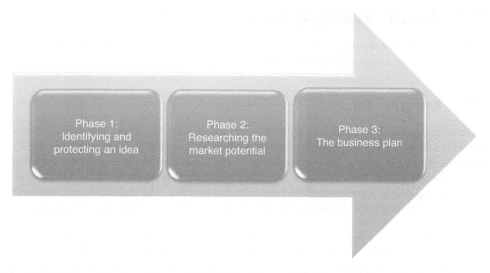

Figure 12.1 Taking a bioscience idea and transforming it into a business proposition can be viewed as a three-phase process involving the identification and protection of an idea, evaluation of the idea in the context of the market and then finally translating the idea into a business proposition in the form of a business plan

misconception is that KT involves a one-way transfer of knowledge from universities to industry. In fact, knowledge transfer is very much a two-way process. It certainly involves the transfer of knowledge, based on the results of research, from universities to industry. However, it also involves the transfer of knowledge, relating to market opportunities and the potential exploitation of research results and ideas, from industrialists to academics. Many universities have enterprise and innovation offices staffed by KT professionals. It is essential that you should be aware of these facilities and individuals from the outset. They are there to help ensure the two-way KT process occurs most effectively and that your ideas are given every opportunity to attract funding and realise their full potential. Take the time to find out more about the KT/Enterprise and Innovation Office at your university. If you have a good idea, they are likely to be very pleased to hear from you!

Exercise

Take a few minutes to list the skills and qualities that you think are needed to make:

(1) a good scientific researcher;

(2) a successful entrepreneur or business person.

Do these roles have any skills and qualities in common?

How many of these skills and qualities do you already have and which do you need to develop?

Keep these skills and qualities in mind as you progress through this chapter. At the end, re-assess yourself to see if what you have learned has changed your skills profile as a potential bioentrepreneur.

12.2 Phase 1 Identifying and protecting an idea

In this section we will consider how you might identify a commercially exploitable idea, and the approaches you can adopt to protect your idea.

12.2.1 Having an idea

Your entrepreneurial journey begins with an idea, and in the biosciences this might involve a research result or the identification of a novel application for a technology. However, you should not be constrained by the subject you study and you should always be on the look out for exploitable ideas in all walks of life. You may decide you would like to try to generate novel ideas from scratch, perhaps in response to a tricky problem you are trying to solve. The techniques and approaches described in Chapter 1 are designed to help you generate creative ideas either as an individual or as part of a team.

You may have an idea that is not based on the results of your own work. This is most likely to happen if you read widely and consult publications beyond your immediate area of interest. A multidisciplinary approach will often give you a novel perspective on your own studies and research, and may lead you to have interesting and imaginative ideas. The story of penicillin nicely illustrates the importance of reading the scientific literature as a source of inspiration for ideas. Alexander Fleming's original observation of inhibition of bacterial growth by a fungus was published in 1929. However, it was not until around 1940 that Howard Florey and Ernst Chain read Fleming's paper, realised the therapeutic potential of the antibacterial agent (penicillin) and isolated enough penicillin for clinical trials. Their results led to a revolution in healthcare and the new age of antibiotics.

12.2.2 How do I know if I have a novel and exploitable idea?

Having identified an idea, you now need to establish its value and novelty. There are a number of sources of information you can consult including the following:

- **Literature sources:** By scanning the literature in your own and related disciplines, you should be able to determine whether your idea is novel and whether it has already been patented or commercially exploited in any way. Examples of relevant literature sources are: academic and industrial peer-reviewed journal articles, conference proceedings, textbooks, book reviews and reports generated by companies such as Ernst & Young (see Section 12.10, Additional resources).

- **Electronic resources:** There are many databases and search tools that you can access including: the US and European patent databases, academic and industrial journal databases (e.g. Sciencedirect and Web of Knowledge) and Google (see Additional resources).

12.2.3 Protecting ideas

12.2.3.1 Good laboratory practice and enterprise

As indicated in Chapter 3, you should keep a thorough record of all of the work undertaken in the laboratory. If you spend some time with a company, for example during an industrial placement, you will probably find that this good practice extends to you dating and signing the work, and your supervisor counter-signing your entry. This will confirm ownership of the results. You will also be encouraged to be very careful about who has access to your laboratory notebooks. This will ensure that your results are secure and not in the public domain. This level of security will be essential should you wish to protect your results through, for example, patenting (Section 12.2.3.4).

12.2.3.2   Publish or protect?

Bioscientists naturally wish to share the outcomes of their work with others but are frequently given conflicting advice regarding the publication and dissemination of their results. On the one hand their careers often depend on the publication of research findings in reputable, peer-reviewed journals. On the other hand, if the results of their research are released into the public domain through publication in journals, or by any other routes, then it will subsequently prove very difficult or impossible to protect the results from exploitation by competitors. If you believe you have an original idea or set of results that you may wish to exploit commercially, you should not consider immediate publication of this information. Instead you should seek advice from your supervisor and the IP (intellectual property) professionals in your institution about the best way to protect your IP rights (IPR). The KT/Enterprise and Innovation Office at your university will be able to put you in touch with an appropriate adviser.

12.2.3.3   Maintaining confidentiality

If you need to discuss results that you may wish to protect with colleagues who are not employees of your university (or industrial placement company), then you should set up a confidentiality agreement before you speak to them. Such agreements are known as Non-Disclosure Agreements (NDAs) or Confidential Disclosure Agreements (CDAs); your university will usually have templates for these which you can customise for your needs (see also Additional resources).

You can protect your IP in a number of ways, and major approaches are outlined in sections 12.2.3.4–12.2.3.6. For a much more detailed consideration of the protection of IP in the context of bioscience, see Byass (2008).

12.2.3.4   Patents

Patents provide a well-known and effective form of protection of IPR that can last for up to 20 years. They are used to protect inventions in the form of products or processes but cannot be used to protect ideas alone. A patent will only be granted if a number of rigorous criteria are met. The invention must: be novel (it must not have been described before or disclosed to others before the date of filing the application); be inventive (not obvious to another expert); have industrial applicability (have some practical use); not be excluded by statute (some things are not considered patentable, for example, musical works, computer programs or immoral inventions e.g. landmines). Patents are only applicable in the countries in which they are filed, although most inventors will wish to apply for protection in a number of countries and this can usually be readily achieved.

12.2.3.5   Copyright

Copyright cannot be used to protect an idea but it can be used to protect the *expression* of an idea. It exists automatically once an original work has been created and it is therefore not necessary to formally register copyright. An original work in the biosciences might be a publication, database or piece of software. If a third party chooses to copy or issue your work without your permission then this will be an infringement of copyright. You can indicate to others that your material is copyright work by adding your name, date and the © symbol. If you wish to prove you were the first to create a work then you should post a copy to yourself or a legal adviser using registered post. Once received, the envelope should be left unopened and kept as a record that, on the date of posting, the work was created by you. Copyright is clearly an inexpensive form of protection but it does not offer the level of protection afforded by patenting. On the other hand, registering a patent may cost thousands of pounds.

12.2.3.6 Trademarks

Like copyright, these do not directly protect an idea; instead they are used to *identify* a product or organisation. Trademarks take the form of graphical illustrations that distinguish the products of one organisation from those of another and are closely associated with brand image and reputation. Once registered, competitors are not permitted to use the mark in relation to their products. The owner of a registered trademark is entitled to use the ® symbol with their mark; good examples are Viagra® and Genentech®.

12.3 Phase 2 Researching the market potential for your idea

In this section we will consider how you can research and evaluate your idea in the context of the market; in other words, how your idea might be exploited in a real-world setting. You need to establish the value of your idea to a prospective customer and the reason(s) why he or she should buy your product or service rather than the product/service offered by your competitors. Remember the aphorism 'the customer is always right'. It is a good rule to bear in mind when identifying customer and market value; you need to know what the customer wants from your product or service and what is already available in the marketplace. Bear in mind that science and technology are highly dynamic sectors and the needs of your target market and consumer base will be constantly changing. Therefore, being able to *predict* market trends is just as important as understanding the current market.

There are clearly many issues that must be addressed as you evaluate the market potential of your idea, and several of these are considered in the following example:

A number of students worked together during a final year group project to develop paper hand tissues containing a reagent that interacts with a marker present in the viruses that cause the common cold. The reagent changes colour when it binds to these viruses, thus providing an early warning of transmission of infectious agents or the likelihood of developing a cold. This would prepare individuals for the onset of cold symptoms and, hopefully, encourage them to avoid contact with others. The students are now thinking of setting up a business, Cold Comfort, *to market this novel approach.*

With regard to customers, there are a number of key questions that the students will need to ask:

(1) Is there a customer need or problem that the idea will address and solve? They need to be very clear about the target market and why the product fills a niche within that market.

(2) Precisely who will be the customers? The students need to know about the types of people (age, sex, ethnicity, income etc.) buying similar products and how these are changing (demographic trends). They will also need to research 'psychographic' factors, i.e. the values, attitudes, interests and lifestyles of potential customers.

(3) What factors will impact upon the relationship with customers? For example, will they expect very high-quality products or exceptional after-sales support? The answers to these questions will help the students see where the true value of their product lies.

Research into the nature and needs of customers is an important part of market analysis that must be conducted during the preparation of a business plan (Section 12.4). A key element of this analysis is the acquisition of competitive intelligence (CI), which can be defined as the process by which information about customers, competitors, products, services etc. is gathered and used for business planning. The students could accumulate this sort of information using questionnaires and other survey techniques. Alternatively, they could access a market research database such as Mintel, to which the majority of universities subscribe, or hire CI professionals who will gather large amounts of data for them (for more information about CI and market research techniques see Wilkinson and Selvaratnam, 2008, and Section 12.10, Additional resources).

A range of additional factors besides customers and competitors will impact on the success or otherwise of the students' business venture, and they need to consider these factors alongside all of the information they accumulate from market research. They decide to do this using two powerful tools known as PEST and SWOT analyses. These techniques can be employed most effectively by constructing grids (Tables 12.1 and 12.2). In PEST (Political, Economic, Social, Technological)

Table 12.1 PEST analysis for *Cold Comfort*

Political	Economic
The UK government has announced a new policy which aims to raise awareness of the effectiveness of hygiene in preventing the spread of disease. In line with this policy, government funding is available, through various bodies, for companies trying to tackle this issue. There are UK and EU regulations and legislation that need to be adhered to when developing new reagents or equipment for the diagnosis of human disease.	UK economy is strong and the medical diagnostics market is experiencing rapid growth. UK government has recently announced a reduction in taxation of medically related health products. During the last five years, once a prototype has been tested, a common route to market for biomedical products has involved acquisition of smaller concerns by large pharmaceutical companies.
Social/cultural	**Technological**
This product is potentially useful to anyone who uses hand tissues and may be particularly popular during the approach of the winter season when the incidence rate of the common cold increases significantly. Older individuals, who are more susceptible to infection, can be expected to benefit most; in their case a route to sale might be through NHS-based prescriptions. Hand-tissue sales figures indicate consumer buying peaks in mid-summer and from November to February. Existing hand-tissue manufacturers already have a strong market presence: breaking into this market may prove difficult.	Currently there are no other companies using this technology in any global market. We have secured European patent rights to use and sell the technology and product. Our product is ready to launch or sell but is manufactured in our specialised site in the UK.

Table 12.2 SWOT analysis for *Cold Comfort*

Strengths	Weaknesses
Our unique selling points are: The product could help significantly decrease the number of colds that grow into more threatening situations, such as pneumonia and flu, due to individuals being able to seek medical treatment at an earlier stage of the infection process. As a method of increasing awareness of disease, the product has the potential to decrease the spread of illness by encouraging people to pay attention to personal hygiene. The underlying technology and product are protected by patent in the EU.	It will be hard to penetrate the current hand-tissue market independently as a new brand. The company will need at least £30000 to produce the first bulk orders. The product will be easily copied by competitors through reverse engineering. This could be a problem if patents aren't granted quickly in the USA and Asia-Pacific region. However, additional patenting is likely to cost thousands of pounds.
Opportunities	**Threats**
Lifestyle trends suggest people are working increased numbers of hours and can't afford to be ill for a long period of time, if at all. The ability to fight an infection earlier and remain at work will be potentially very beneficial to many. There are significant opportunities to develop partnerships with large, overseas pharmaceutical companies. These companies have the capacity to distribute products to global regions which are in urgent need of better health diagnostics because illness in these countries can quickly develop into serious health complications through lack of medicines.	With the increased effectiveness and range of cold- and flu-symptom relief medicines available in developed countries, there may not be a large enough market demand for this diagnostic product. The ability to produce the product depends upon the highly specialised knowledge of key team members. Common cold variants could easily mutate and render our biomarker useless in future months or years.

analysis, as many external factors as possible that could possibly have an impact on the business are considered. SWOT analysis involves a comprehensive consideration of the Strengths, Weaknesses, Opportunities and Threats involved in a business venture. When the students consider the combined results of their PEST and SWOT analyses they should be in a much better position to assess the viability of their business idea.

Exercise

Imagine the students now ask for your opinion on whether or not they should go ahead with the *Cold Comfort* business venture. Given the results of the PEST and SWOT analyses, what would you advise at this stage?

12.4   Phase 3   Setting out your ideas and goals – the business plan

You have identified an idea and a market opportunity. The next stage is to express these clearly in a business plan. The business plan is a very important document that should contain a set of goals, the reasons why they are considered attainable, and the plan for achieving these goals. It should also contain information about the individual or team behind the plan. You will use the business plan to 'sell' your ideas to potential investors. It will hopefully help you to gain access to funding, and other resources and knowledge in support of your business venture. The plan will also provide an opportunity for you to work out how you will manage, develop and grow your business.

You are probably familiar with the popular television show *Dragons' Den*. The Dragons ask lots of tricky questions and they expect the would-be entrepreneur's answers to reveal whether or not he or she has compiled a thorough business plan. One effective way to construct a business plan is therefore to ask yourself the questions you think a potential investor ('Dragon') might ask. If you are to avoid hearing 'I'm out' then, at the very least, you will need to know the answers to the following key questions:

(1) Is your idea robust and can you prove it will work?

(2) How will you protect your idea?

(3) Why is your product or service unique compared to what competitors can offer?

(4) How does your product or service fit into the current market?

(5) How will your business make money?

(6) What are the risks and how will you manage them?

(7) Why should investors support your idea rather than the ideas of others?

(8) What level of investment (funding and expertise) do you need?

Exercise

Read the Case study that describes an interview with Dr Curtis Dobson of the company Ai2, and think about how you would have answered the above questions if you had been a founder of Ai2. Do you have a business idea of your own? If so, do you feel you have satisfactory answers to some or all of these questions?

12.4.1   Case study   Interview with Curtis Dobson, founder and CEO of Ai2

Ai2, a spinout company founded in 2005 by Dr Curtis Dobson of the University of Manchester, produces anti-infective medical device coatings. These novel, human protein-derived coatings safely and cost-effectively inhibit the growth of bacteria, fungi and viruses on commonly used medical devices such as urinary catheters, stents and wound dressings. They have been proven to be extremely effective in preventing infection.

You founded Ai2 in 2005 – did the commercial idea arise from scientific research, or were you searching for a solution to a problem that had already been identified?

The technology goes back to 1998 when I first did the work in collaboration with someone in Cincinnati who I met at a conference in Amsterdam. The collaboration was around developing a research tool for use in Alzheimer's disease. We discovered a peptide that was derived from a human sequence, effectively a bit of the human proteome, which was powerfully anti-infective; so we were struck that this was something that could be commercially valuable. It was very much that the science came first and we kept looking around for something to do with that, rather than thinking 'Here's an unmet need, let's find a solution to it.'

Was the potential commercial value immediately obvious?

Not really; at the time we were using viruses as a model and our initial thought was this could be great as an anti-HIV therapy. There was some interest in the idea, but it is a very crowded marketplace and there are lots of companies developing therapies for HIV infection . . . and the timescales are vast, for a small company to develop a therapeutic it's quite a big undertaking. It's also very risky . . . So it was about 2003 when I'd spoken to colleagues here (at the University of Manchester) about other potential applications of the technology that we started to look at the anti-bacterial and anti-fungal effects of this technology. We knew that there was a problem with infection in certain medical devices, contact lenses in particular, and we found (from a few proof-of-concept experiments), quite interestingly, that the peptide actually self-adhered to plastics, so that led us to look at this further, but it was quite a journey from the original research, through to therapeutics, through to medical devices.

Why were you interested in the business angle of bioscience?

There was quite a career incentive – at the time I was a non-tenured post-doc and this was a way of carrying on the science . . . with a potential longer-term incentive of a windfall down the line . . . On the other hand, it's an opportunity for the science to actually make a difference.

Did you follow a particular path?

We started out with an IP search, checking that no one had already patented our molecules. There was no point filing a patent until there was funding to develop that patent, because it has to be finalised within a year of filing; so next, we secured POP (proof of principle) funding, which was a few tens of thousands, to fund someone in the lab for a year so that we could fully understand that the invention worked. Then we filed the patent, but needed more funding to take it to next stage; so we started to talk to potential licensees, end-users of the technology. At that stage we also started putting together a more formal business plan and creating a team of people to put the business plan into practice. There was myself, UMIP [University of Manchester Intellectual Property] and investors ... and we put a formal board of directors together. We employ a chairman and chief executive with experience in medical devices: someone who has run a large public company before and understands the market and who has a lot of contacts.

What risks and challenges did you face?

On a personal level, if the business hadn't been successful early on it could have left me without an academic or commercial career.

There is a difficulty engaging with large corporations, and there is a risk that they won't be interested, and this is quite a difficult process to manage. We did do market research, but we already had good contacts with the medical devices sector through the university. Since then, we have looked at other sectors, and employ a business development manager to cold call almost – to sit by the phone and send off data packs. As a result, we now have a couple of other licence opportunities that are reasonably close to being secured in a couple of other areas. Ultimately it is about making those connections and talking to the people directly involved.

Is Ai2 a success story, and where are you now?

I'd like to think it was! I think I've progressed much further than I would have done had I just been publishing papers and getting in grants. It has helped secure my career and promoted my science to a much wider audience. It helped by being at this university, where there has been increasing emphasis on commercialisation and translation of research.

There are still technical and regulatory hurdles to get through, but once the product is there, from the scientific point of view, it will be quite a prominent product. Any kind of medical device product might well feature this type of peptide. We expect to be profit-making within two years, which is why we need investment. One of the difficulties for any biotech company, particularly therapeutics, is that you've got to find investment even though it will not be profit-making for at least a decade.

What advice would you give to would-be entrepreneurs?

For 95% of students, the most obvious route to follow is to get training and experience in the right areas, but, potentially, PhD students can start companies. There is a very obvious example, Google, of course – quite a good success story! What you need to do is to identify an unmet need and have a novel and inventive solution to that need. The next step is to check the patent databases to make sure that no one else has thought of it, and then it might be developable as a business. Although Ai2 started with the technology, it can be done the other way round: you can look at a problem and search for a solution to it. Recognising the unmet need is key: science and medical websites are full of unmet needs – infection is an obvious one. Don't necessarily go in for the glamorous ones like curing AIDS. Catheter infection is certainly not glamorous, but it is a big problem! Reading and taking any opportunity to talk to people will help you see where the problems are.

People have a view that the commercial world is in some way in the opposite direction to the science, it's not at all; it actually enhances it.

12.4.2 Bioscience business models

If you are working within a university and wish to commercialise a product/technology, you may decide to follow the route taken by Dr Curtis Dobson and establish a *spinout* company. A spinout company splits off from a parent company or organisation (in this case a university) to become an independent business. Universities are frequently keen to support spinouts, and will often offer generous support as part of their knowledge transfer programmes. There are other options for bioscience business. For example you might decide to license your IP, sell the technology you have developed or form a strategic alliance with an industrial partner. Price and Sarmiento (2008) provide further information about alternative models for bioscience enterprise. However, let us assume that you wish to go ahead and establish your own business.

12.4.3   General business-plan structure

There are many ways in which to structure and write a business plan; the one you choose will be dependent upon factors such as the nature of the idea (a product or service) and the nature of the investor who will be reading your plan. However, every good business plan should include a number of fundamental components, and these are summarised in the following sections. Some of the specific issues associated with bioscience business plans are considered in Section 12.4.4, and you will find more detailed overviews of the business-plan structure elsewhere (for example, see Ferguson (2008)).

12.4.3.1   Introduction

An introductory section should consist of the following:

- Cover Page
- List of Contents
- Executive Summary
- Skills profiles for the people involved.

The Cover Page should include the title of your business and contact details for the key people involved. A Contents List will help an investor find what he or she is looking for quickly. This is particularly important for science-based business plans, which can be very elaborate; if potential investors are able to navigate the different sections easily then they are less likely to miss important parts of the plan. The Executive Summary is a synopsis of the entire business plan and should include key elements such as your mission (purpose) and vision (where your business will go) statements, and the level of investment you require; the Executive Summary should be written on a single side of A4 paper. Lastly, you need to introduce the people behind the product or service: provide backgrounds for team members and indicate the skills each brings to the business.

12.4.3.2   The opportunity, market and strategy

The fundamental points to cover in this section are:

- the opportunity and idea;
- an indication of potential markets, with market analysis;
- outline of your objectives and a strategy that indicates how you will meet these objectives.

First, you should describe the opportunity and indicate how your idea will exploit that opportunity. Crucially you must define the value of your product or service to the customer; always remember that what seems like a valuable and useful idea to you might not be perceived as valuable and useful by others. You should also indicate here whether you have protected your idea, and the level of protection you have achieved.

Next you need to provide details about the potential market(s); this is a vitally important section of the business plan. The results of your SWOT and PEST analyses (Section 12.3), along with any conclusions and predictions drawn from them, will provide much of the information you need for this part of the business plan.

Finally in this section, you must develop an operational plan. This should detail your objectives and any strategies you intend to employ in order to meet these objectives; you will need to take into account and integrate the results of your market analysis as you formulate the operational plan. Key terms to include in this section are: objective, strategy, tactic and self-assessment. For *Cold Comfort*, the business considered by the university students in Section 12.3, these terms might be used as follows:

Objective: the annual turnover for the second year of operation of *Cold Comfort* will be £2 million.

Strategy: to meet this objective the company aims to form strategic partnerships with international hand-tissue and hygiene product manufacturers and an assortment of health institutes.

Tactic: in the first quarter of Year 2, all necessary resources and expenditure will be dedicated to the international section of the company so that it can develop the necessary strategic partnerships.

Self-assessment: weekly review meetings will be held with investors, directors and heads of department to ensure that the company is on track to meet its objective.

An operational plan is commonly summarised in a Gantt chart. This is a very effective way of making a potential investor aware of the company's strategy over a given period of time. In Figure 12.2 you can see how *Cold Comfort* planned their first year of operation.

12.4.3.3 Investors, finances, risks and needs

This section of the business plan should include a detailed overview of the financial components of your business, what you will need from an investor and the risks you have identified. The financial elements are usually worked out, for the most part, using assumptions: they are therefore *estimates*. When working with assumptions and estimates it is important to be entirely honest and transparent, particularly because there might be a complex web of assumptions, such as those related to the market, competitors, available resources (raw materials), economic situation and other largely unpredictable factors.

Types of investor
Before writing the financial section it is important to consider the *type* of investor you will be presenting the business plan to. Common options are: banks, business angels, venture capitalists and funding bodies that distribute grants and awards. These potential investors differ markedly:

• Banks usually offer loans to finance businesses, and represent a relatively predictable form of finance. They charge interest on loans but are unlikely to want a share in the business or any say in how it is run. Interest rates for loans can vary depending on individual agreements.

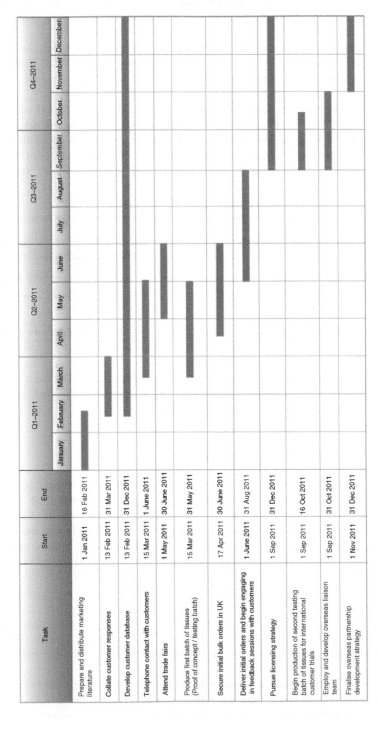

Figure 12.2 A Gantt chart for the first year of operation of *Cold Comfort* (for further information about Gantt charts see Additional resources)

- Business angels are individuals or small groups that have successful businesses themselves and have the money and the time to invest in new ventures. They contribute expertise as well as money, and usually demand a significant share of the business.

- Venture capitalists are usually individuals that deal with investments totalling millions of pounds which they use to buy a share in a new business. A financial return is reaped as the business grows, and is sold on (in as short a time as possible for greatest return).

- Grants and awards, for example from charities and foundations, are sometimes available for business start-ups. These usually have strict criteria which you, the applicant, must carefully address.

For additional advice about funding your ideas see Baynes and Pymer (2008). Your university's KT/Enterprise and Innovation Office should also be able to provide detailed information about potential funding sources for business ventures in the biosciences and elsewhere.

Once you have decided on the type of investor you will pitch your business proposition to (you may decide to seek funding from more than one type of source), you will need to build the financial section of your business plan. This section of business plans generally consists of the following core parts:

- financial forecasts

- risks analysis

- investment requirements.

Financial forecasts

Your financial forecast should include several core aspects such as a sales forecast, cashflow summary (flow of cash in and out of a business) and an account of 'profit and loss'. When making sales forecasts you need to be as realistic as possible. For example, it is quite normal, especially for bioscience companies, to make a loss in the initial stages whilst elements of legislation and testing are pursued and completed. When forecasting, remember to explain the reasons behind your forecasts so that an investor can see the logic and reasoning you have used; this is perhaps even more important than the numbers and estimates! The most common way in which you might present your overall finances to an investor is through the use of a profit and loss account. This is essentially a statement of all income and expenses over a given period of time and includes gross and net profit or loss. For further information about profit and loss accounts, with examples, see the Business Link website listed in Section 12.10, Additional resources.

Risks analysis

Next you need to demonstrate that you have done all you can to identify the risks associated with your business proposition, and you must indicate how you will manage these risks. At this stage it is useful to refer to the results of SWOT and/or PEST analyses (Section 12.3). Once you have listed all the potential risks it is important to grade these in terms both of *impact* and the *probability of occurrence*. Impact refers to how severe the consequences will be to the success of the business. This process can be simplified, and the risks presented clearly, by constructing a grid (Table 12.3).

Worked example

The student directors of *Cold Comfort* consider the results of their SWOT and PEST analyses (Section 12.3) and assess the risks associated with the establishment and expansion of their business as follows:

Table 12.3   Risk quadrant indicating impact and likelihood of risk occurring

<table>
<tr>
<td rowspan="2" style="text-align:center">IMPACT</td>
<td style="text-align:center">High</td>
<td>• Common cold virus mutates to render product useless.
• Competitors avoid patent restrictions by creating highly effective product that is similar, but not identical, to *Cold Comfort*'s.</td>
<td>• No US or Asian patents.
• Competitors have strong market presence.</td>
</tr>
<tr>
<td style="text-align:center">Low</td>
<td>• Investors withdraw funding.
• Insufficient market demand for product.</td>
<td>• Energy and transportation costs increase.
• *Cold Comfort* is dependent upon the expertise of a small number of individuals.</td>
</tr>
<tr>
<td></td>
<td></td>
<td style="text-align:center">Low</td>
<td style="text-align:center">High</td>
</tr>
<tr>
<td colspan="4" style="text-align:center">PROBABILITY OF OCCURRENCE</td>
</tr>
</table>

If you use this approach you need to justify classification of risks in this way, and indicate to investors how you plan to deal with the threats. For example, the presentation of risks in the above grid could be explained and elaborated upon as follows:

Likely to occur with high impact

These risks require immediate attention.

No US or Asian patents: the company directors are currently doing everything they can to ensure patents are in place in the USA and Asia in the near future.

Competitors' strong market presence: to counter the strong market presence of competitors, the directors are contemplating an aggressive advertising campaign. In the longer term they will consider licensing the product to competitors.

Unlikely to occur but would have a high impact

Virus may mutate: the epitope that interacts with the reagent is highly conserved and thought unlikely to be modified markedly by mutation of the encoding gene.

Competitors create similar product: the product is novel so it seems unlikely that competitors will create a related product that does not violate patent law.

Likely to occur but would have a low impact

Energy and transportation costs increase markedly: seems inevitable but the business plan has allowed for this, so impact should be minimal.

Dependence on small number of experts: the company intends to employ additional highly skilled individuals in the near future and is building collaborative links with other organisations prepared to share expertise.

Unlikely to occur and would have a low impact

Investors withdraw funding: the company has secured guaranteed funding for three years. The directors are confident that prospects are currently excellent and there is no reason for current investors to lose interest and withdraw longer-term funding. However, if they did, it seems likely that the company will attract funding from other sources.

Insufficient market demand for product: the results of market research (Section 12.3) suggest that there will be a strong demand for the *Cold Comfort* product.

What you would like from an investor and what the investor can expect to gain

In completing this section of the business plan you need to outline exactly what you are asking for in terms of an investment and what you will give for that investment. Typically, new businesses require financial investment, or expertise and guidance, or both. In return for this investment you, the entrepreneur, will usually be expected to offer a share of your business *(equity)* which is intended to provide the investor with a substantial positive return on their investment. At this stage remember that it is not all about what you can offer: as an entrepreneur you must consider if the investor can offer what *you* need. An ability to negotiate effectively (see Additional resources) at this stage is crucial if you are to attract the best investment possible. A key part of an investment, which will feature in these negotiations, is known as the *exit strategy*. This is a term used to describe how an investor will recoup their investment through the sale of their part of the business. You, the entrepreneur, are also likely to need an exit strategy; for example, a plan for how you will sell the business once it is profitable.

Concluding your business plan

In concluding the business plan it can be helpful to start by writing a short summary of each of the core sections of the plan described in this chapter. Then try to condense all of these summaries into a single paragraph that concisely and effectively conveys: your idea; how it fits into a market; what you estimate the business to be worth at various time points in the future; and what you will need from the investor. Additionally, this final section offers you an opportunity to underline the fundamental points you made in the executive summary and throughout the business plan. Rather than simply repeating these messages you should, as appropriate, elaborate upon them and include any further details that you think important or potentially attractive to an investor.

When writing your business plan you should bear in mind that information contained within it may be of a sensitive nature. This does not mean you should exclude such information from the plan. Instead, ensure that you take legal advice and include within the business plan a formal understanding of confidentiality such as a non-disclosure agreement (see the Additional resources section for information on where to find an NDA template). This will ensure that whoever reads the business plan and signs the NDA (for example, a potential investor) will not be permitted to share details contained within the plan with others.

Appendices to the plan can also be useful. They allow you to include supplementary, detailed information that is relevant to the business but that may not be required to enable an investor to decide whether or not he or she should invest in your idea. Examples include curriculum vitae (CVs) for staff; patent documents; detailed reports, for example, market analysis; detailed explanations and specifications of the technology; and any other supporting data such as financial records (Ferguson, 2008).

The Additional resources section at the end of this chapter contains a number of links to very useful websites that provide more detailed information and advice about compiling a business plan.

12.4.4 The bioscience business plan

All business plans need to be tailored to the appropriate audience. Business ideas that stem from the sciences are frequently viewed as high-risk affairs, and you must do all you can to allay the fears of potential investors. In particular, when compiling a bioscience business plan, there are a number of specific factors that must be taken into account including the following:

(1) **Knowledge and skills dependence:** bioscience businesses are often dependent on the highly specialised skills of a small number of individuals. Their knowledge and expertise may have been acquired during a long career of research in the laboratory and may not be easily passed on to others. Business plans that depend on key individuals in this way should therefore incorporate a strategy for how the business will be sustained if these individuals need to leave the company.

(2) **Long lead or development time:** bioscience businesses can take a particularly long time to bring a product to the market. Once again this is exemplified by the pharmaceutical industry: it is estimated that, typically, it will take 12 to 15 years for a new drug to be developed from idea to market (Masia, 2006). You will need to make plain the time required to develop your idea to the stage where you are offering a product for sale, and you will need to indicate how your business can be sustained during any lengthy development period.

(3) **Legislation:** new regulations and policies can have a severe impact on a bioscience business. This is particularly the case in the pharmaceutical industry where, for example, regulations for drugs trials involving humans or animals are regularly revised. You need to ensure you are fully aware of any new legislation that may be introduced that is likely to affect your business activities.

(4) **Patents:** is your idea actually patentable? Elements of nature, such as naturally occurring processes and molecules cannot be patented; however, using them in novel ways can be patented. For a basic overview of patents for life science companies see the Additional resources section.

(5) **Types of investor:** you need to think carefully about the type of investor you will wish to approach (see Section 12.4.3.3). Science-based businesses often wish to attract investors who can provide expertise, and there may be a lengthy development time before the product reaches market. A bank loan that comes with no specialist expertise but with high interest rates may therefore be far less preferable to, for example, a combination of funding and support from a business angel, with sector expertise, and an enterprise award from a charitable foundation.

12.5 Communicating your business – the 'Pitch'

'I'm in!' These are the words you want to hear after pitching your business proposition. Unfortunately, in reality, many business pitches fail to accurately and concisely communicate either the problem or the solution, and the would-be entrepreneur therefore loses out on investment. In the case of bioscience business start-ups this is often because it can be very hard to condense highly complex problems and solutions into a few sentences. However, it is not impossible: the concise communication of complex concepts is a skill that can be learned. For example, you can practise and develop the delivery of an 'elevator pitch'. This term relates to a hypothetical situation where you might find yourself in an elevator (lift) with a potential investor. Typically you will have up to one minute to tell the investor about your business idea; i.e. you will have the opportunity to make a one-minute elevator pitch. You need to practise how you will communicate all of the essential information about your business idea in that short period of time. There are several factors you should consider:

- Begin the pitch with an attention-catching statement.

- Describe what the core problem or opportunity is and how you have solved the problem or exploited the opportunity.

- Stick to using numbers and facts, wherever possible, rather than estimates.

- Include any highly significant evidence such as patents, commitments to buy or preliminary data.

- The way in which you present your pitch needs to convey desirable characteristics about you and your team such as: enthusiasm, confidence, honesty, integrity and a willingness to cooperate.

- End the pitch by outlining the level of investment you need.

Once you have decided on an elevator pitch for your business, practise on family and friends! See Chapter 11 for further advice on developing your communication skills.

Exercise

Imagine you are CEO for the company Ai2 (Case study, Section 12.4.1). You suddenly have the opportunity to make a one-minute pitch to a group of potential investors you meet in the lobby of a hotel. Refer to the Case study and decide what you would say about the company during a pitch of 60 seconds.

12.6 Concluding comments

This chapter has taken you on a journey: we have demonstrated how you can start with an idea and transform it into an attractive and valued business proposition. The chapter should also have raised your awareness of the potential rewards and the considerable risks associated with bioscience business. However, many of you will pursue a career that involves little, if any, bioscience. Hopefully the approaches described here will encourage you to adopt an enterprising attitude in any of the workplace settings you encounter – good luck!

12.7 How you can achieve your enterprising and entrepreneurial potential

- Unlock your creative potential (see Chapter 1).

- Look out for 'unmet needs' in the biosciences and elsewhere.

- **Record** your thoughts and ideas.

- Understand how to protect your ideas.

- Get advice from the professionals in your university's KT/Enterprise and Innovation Office.

- Know how to carry out market research and write a business plan.

- Practise your 'pitch' at every opportunity.

- Get involved in extracurricular projects that involve enterprise.

- Obtain first-hand experience of enterprise during a work placement year.

- Consider registering for a post-graduate business qualification.

12.8 Tutor notes

A number of organisations support academics who wish to teach enterprise and entrepreneurial skills.

The UK Centre for Bioscience, Higher Education Academy, provides a range of resources in support of enterprise teaching (www.bioscience.heacademy.ac.uk/resources/enterprise.aspx) including an Enterprise Audit Tool and an Enterprise Skills Matrix that together help staff identify and enhance enterprise elements within a degree programme. The Centre for Bioscience also provides a number of web pages that help university staff address entrepreneurship in the context of bioscience (www.bioscience.heacademy.ac.uk/resources/entrepreneurship/index.aspx).

The National Council for Graduate Entrepreneurship (www.ncge.org.uk) aims to raise the profile of entrepreneurship in UK higher education. It rewards those who support student and graduate entrepreneurship in Higher Education through its National Enterprise Educator Awards.

The Wellcome Trust provides a group activity that allows students to practise their entrepreneurial skills in a tutorial or mini-project setting. It encourages them to work together as a team, and to analyse, evaluate and critically review information and evidence so they can present a case for the development of a product (www.wellcome.ac.uk/Education-resources/Teaching-and-education/Big-Picture/All-issues/Drug-development/Student-activity/index.htm).

Finally, the Biotechnology and Biological Sciences Research Council (www.bbsrc.ac.uk/home/home.aspx) runs a series of Commercialisation and Development programmes including the very popular Young Entrepreneurs Scheme (YES). YES is open to anyone working as a researcher in the biosciences, although it is aimed primarily at individuals in the early stages of their career (in practice this usually means PhD students or individuals on their first or second post-doc appointment). The scheme aims to raise the entrepreneurial awareness of participants who present their plans for hypothetical businesses to real entrepreneurs, financiers and industrialists. You may wish to encourage teams of post-grads or post-docs from your institution to enter this annual competition.

12.9 References

Baynes, D. and Pymer, L. (2008) Funding your ideas. In *Enterprise for Life Scientists: Developing Innovation and Entrepreneurship in the Biosciences*, Ed. Adams, D.J. and Sparrow, J.C. Bloxham: Scion, pp. 181–200.

BBSRC (2010) *The Age of Bioscience: Strategic Plan 2010–2015*. Swindon: Biotechnology and Biological Sciences Research Council (BBSRC). Available at: www.bbsrc.ac.uk/web/FILES/Publications/strategic_plan_2010–2015.pdf. Accessed January 2011.

BIO (2008) *Guide to Biotechnology 2008*. Washington, DC: Biotechnology Industry Organization. Available at: http://bio.org/speeches/pubs/er/BiotechGuide2008.pdf. Accessed January 2011.

Booth, S. and Armour, K. (2008) Knowledge and technology transfer. In *Enterprise for Life Scientists: Developing Innovation and Entrepreneurship in the Biosciences*, Ed. Adams, D.J. and Sparrow, J.C. Bloxham: Scion, pp. 1–24.

Byass, L. (2008) Protecting ideas. In *Enterprise for Life Scientists: Developing Innovation and Entrepreneurship in the Biosciences*, Ed. Adams, D.J. and Sparrow, J.C. Bloxham: Scion, pp. 52–80.

Ferguson, A. (2008) The role of the business plan. In *Enterprise for Life Scientists: Developing Innovation and Entrepreneurship in the Biosciences*, Ed. Adams, D.J. and Sparrow, J.C. Bloxham: Scion, pp. 150–180.

Masia, N. (2006) The cost of developing a new drug. In *Focus on: Intellectual Property Rights*, Ed. Clack, G. Washington, DC: US Department of State, pp. 82–83. Available from: www.america.gov/publications/books/ipr.html. Accessed April 2011.

Price, A. and Sarmiento, T. (2008) Starting up a business. In *Enterprise for Life Scientists: Developing Innovation and Entrepreneurship in the Biosciences*, Ed. Adams, D.J. and Sparrow, J.C. Bloxham: Scion, pp. 127–149.

Wilkinson, D. and Selvaratnam, A. (2008) Researching ideas. In *Enterprise for Life Scientists: Developing Innovation and Entrepreneurship in the Biosciences*, Ed. Adams, D.J. and Sparrow, J.C. Bloxham: Scion, pp. 81–102.

12.10 Additional resources (accessed April 2011)

12.10.1 General resources

Biotechnology and Biological Sciences Research Council. www.bbsrc.ac.uk/.
Bioindustry Association. www.bioindustry.org.
Department for Business, Innovation and Skills. www.bis.gov.uk/.
Biotechnology Industry Organization. www.bio.org/.
Business Link. www.businesslink.gov.uk.

12.10.2 Phase 1

Ernst & Young, Biotechnology section: www.ey.com/UK/en/Industries/Life-Sciences/Biotechnology_Overview.
A summary of skills associated with entrepreneurship: www.bioscience.heacademy.ac.uk/resources/entrepreneurship/skills.aspx.
Key websites when searching for academic and industrial journal articles:
  Web of Science. www.isiknowledge.com.
  ScienceDirect. www.sciencedirect.com.
  PubMed. www.ncbi.nlm.nih.gov/pubmed.
  ProQuest. www.proquest.com.
  MyAthens: see your university's personal subscription website.
Patent searching and advice:
  European patent databases: www.espacenet.com.
  Google patent search: www.google.com/patents.
  United States Patent and Trademark Office: www.uspto.gov/.
  Basic patent advice for life science companies: www.pli.edu/public/booksamples/19608_sample5.pdf.
Confidentiality agreement information and examples:
  Business Link: www.businesslink.gov.uk/bdotg/action/layer?topicId=1074415494.
  Example 'short' template: www.articlecity.com/articles/legal/article_232.shtml.

12.10.3 Phase 2

Market research resources:
  MINTEL. A useful source of market and competitor intelligence. www.mintel.com (or via your Athens institution user account).

Worldwide data on market research: www.marketresearch.com; http://managementhelp.org/mrktng/
mk_rsrch/mk_rsrch.htm.
Ernst & Young, Biotechnology section: www.ey.com/UK/en/Industries/Life-Sciences/Biotechnology_
Overview.
Information and templates for SWOT and PEST analyses: www.mindtools.com; www.quickmba.com/strategy/
swot/; www.businessballs.com.

12.10.4   Phase 3

Business plan:
How to prepare a business plan: www.businesslink.gov.uk/bdotg/action/layer?topicId=1073869162.
Business plan examples and financial tools: www.bplans.com.
The Business Link website is a comprehensive knowledge base for individuals wishing to start a business.
www.businesslink.gov.uk.
GANTT charts – Microsoft guide: http://office.microsoft.com/en-us/excel-help/create-a-gantt-chart-in-excel-
HA001034605.aspx.
Negotiation skills:
Tips and literature: www.pon.harvard.edu/free-reports/.
Hints and tips: www.businessballs.com/negotiation.htm.
Test yourself using this resource from Heriot-Watt University, Edinburgh: www.ebsglobal.net/programmes/
negotiation-quiz.
British Business Angels Association: www.bbaa.org.uk/; they are also members of the World Business Angels
Association (www.wbaa.biz).
The perfect Pitch? Dragons' Den: www.bbc.co.uk/dragonsden/entrepreneurs/sharonwright.shtml.
Financing a business venture: http://biotech.about.com/od/financing/tp/Financing.htm.

Appendix

Animal welfare legislation in the United States, United Kingdom and Europe

Animal Welfare Act (1966) USA	Animal (Scientific Procedures) Act (1986) United Kingdom	EU Directive 86/609/ EEC (1986) European Union	EU Directive 2010/ 63/EU European Union (effective from Jan 2013)
• Covers warm-blooded animals except farm animals, birds, rodents and fish • Unnecessary duplication of experiments on protected animals prohibited • Pain, suffering or distress to be minimised • Administration of anaesthesia and/or analgesia where appropriate to minimise pain, suffering or distress • Animals experiencing chronic pain or distress that cannot be alleviated to be painlessly euthanised as soon as possible	• Covers all living vertebrates (except man) from defined stages of gestation, and any invertebrate of the species *Octopus vulgaris* • Use of higher-order primates prohibited • Use of cats, dogs, horses or non-human primates requires special permission • Use of endangered species prohibited • Covers breeding, holding or use of protected animals • All animals to be obtained from licensed breeders • Provides authority for specific programmes of work	• Covers all non-human vertebrates excluding fetal or embryonic forms • Use of endangered species prohibited • Wild-caught animals only to be used if no alternative • Covers breeding, holding or use of protected animals • Provides authority for specific programmes of work • Procedures can only be undertaken for defined, permissible purposes • Can only be undertaken if no alternative to animal use • Must consider principles of the 3Rs	• Covers all living vertebrates (except man) from defined stages of gestation, and cephalopods • Only permits use of animals (including for isolated tissue studies) bred specifically for scientific research • Use of endangered species prohibited unless scientifically justified and no alternative • Prohibits use of wild-caught animals • Use of great apes prohibited

Effective Learning in the Life Sciences: How Students Can Achieve Their Full Potential, First Edition.
Edited by David J. Adams.
© 2011 John Wiley & Sons, Ltd. Published 2011 by John Wiley & Sons, Ltd.

Animal Welfare Act (1966) USA	Animal (Scientific Procedures) Act (1986) United Kingdom	EU Directive 86/609/ EEC (1986) European Union	EU Directive 2010/ 63/EU European Union (effective from Jan 2013)
• Surgical procedures to be undertaken in dedicated facilities, with appropriate provision of pre- and post-operative care • No re-use of protected animals for more than one major operative procedure unless scientifically justified • Minimum standards of housing and care for protected animals • Institutions must establish animal care and use committee (IACUC) • IACUC to inspect facilities, review and approve studies involving protected animals • Investigators to consider alternatives to use of protected animals and principles of 3Rs in licence application • Veterinary care available at all times • Breeders and research facilities licensed by US Department of Agriculture	• Procedures can only be undertaken for defined, permissible purposes • Can only be undertaken if no alternative to animal use • Benefits to science must outweigh potential harms to protected animals • Must apply principles of the 3Rs • Must use species of lowest sentience • Pain, suffering or distress to be minimised • Administration of anaesthesia and/or analgesia where appropriate to minimise pain, suffering or distress • Use of neuromuscular blocking agents without anaesthesia prohibited • All animals to be culled (unless special permission obtained) by an appropriate (Schedule 1) method at the end of a regulated procedure • No continued use or re-use of animals without permission • Licensing of facilities, programmes of work and individuals • Individuals must have appropriate skills, education and training	• Pain, suffering or distress to be minimised • Administration of anaesthesia and/or analgesia where appropriate to minimise pain, suffering or distress • All animals to be culled at the end of a regulated procedure if in lasting pain or distress • Restrictions on re-use of animals • Licensing of facilities and prior authorisation of programmes of work • Individuals must have appropriate skills, education and training • Recommended minimum standards of housing and care for protected animals • Veterinary surgeon required to provide medical advice, and individuals responsible for day-to-day animal care and welfare • Designated authority to ensure compliance	• Non-human primates only used for studies of life-threatening clinical conditions in humans and where no alternative • Any non-human primates must be at least of an F2 generation – the offspring of animals bred in captivity • Covers breeding, holding or use of protected animals • All animals to be obtained from licensed breeders • Provides authority for specific programmes of work • Procedures can only be undertaken for defined, permissible purposes • Defines the severity classification of individual procedures • Retrospective review of procedures involving non-human primates or those classed as severe

Animal Welfare Act (1966) USA	Animal (Scientific Procedures) Act (1986) United Kingdom	EU Directive 86/609/ EEC (1986) European Union	EU Directive 2010/ 63/EU European Union *(effective from Jan 2013)*
• All personnel to be appropriately trained	• Minimum standards of housing and care for protected animals • Named veterinary surgeon required to provide medical advice, and named individuals responsible for day-to-day animal care and welfare • All programmes of work to be reviewed and approved by Institutional Ethical Review Procedures (from 1997) • Regulated and monitored by Home Office Inspectorate		• Can only be undertaken if no alternative to animal use • Benefits to science must outweigh potential harms to protected animals • Must apply principles of the 3Rs • Requires a harm–benefit analysis prior to authorisation • Must use species of lowest sentience • Pain, suffering or distress to be minimised • Administration of anaesthesia and/or analgesia where appropriate to minimise pain, suffering or distress • Defines appropriate methods of humane euthanasia • Restricts the re-use of animals • Licensing of facilities and programmes of work • Individuals must have appropriate skills, education and training and be able to demonstrate competence

Animal Welfare Act (1966) USA	Animal (Scientific Procedures) Act (1986) United Kingdom	EU Directive 86/609/ EEC (1986) European Union	EU Directive 2010/ 63/EU European Union *(effective from Jan 2013)*
			• Mandatory minimum standards of housing and care for protected animals • Institutions to employ veterinary surgeon to provide medical advice, and individuals responsible for day-to-day animal care and welfare • Institutions to establish an animal welfare body to monitor and advise on animal welfare and application of the 3Rs. • Regulated and monitored by a national authority

Index

Effective Learning in the Life Sciences: How Students Can Achieve Their Full Potential, First Edition.
Edited by David J. Adams.
© 2011 John Wiley & Sons, Ltd. Published 2011 by John Wiley & Sons, Ltd.

THE PERIODIC TABLE

G = Noble gas
N = Nonmetals
The rest are metals

Key: Element Name — Atomic No. — Symbol — Atomic weight

Name	No.	Symbol	Atomic weight	Type
Hydrogen	1	H	1.0079	
Helium	2	He	4.0026	G
Lithium	3	Li	6.941	
Beryllium	4	Be	9.0122	
Boron	5	B	−0.81	
Carbon	6	C	12.011	N
Nitrogen	7	N	14.007	N
Oxygen	8	O	15.999	N
Fluorine	17	F	18.998	N
Neon	1	Ne	20.180	G
Sodium	11	Na	22.990	
Magnesium	12	Mg	24.305	
Aluminum	13	Al	26.98	
Silicon	14	Si	28.086	N
Phosphorus	15	P	30.974	N
Sulfur	16	S	32.065	N
Chlorine	17	Cl	35.453	N
Argon	18	Ar	39.948	G
Potassium	19	K	39.098	
Calcium	20	Ca	40.078	
Scandium	21	Sc	44.96	
Titanium	22	Ti	47.867	
Vanadium	23	V	50.942	
Chromium	24	Cr	51.996	
Manganese	25	Mn	54.94	
Iron	26	Fe	55.845	
Cobalt	27	Co	58.933	
Nickel	28	Ni	58.693	
Copper	29	Cu	63.546	
Zinc	30	Zn	65.39	
Gallium	31	Ga	69.72	
Germanium	32	Ge	72.64	
Arsenic	33	As	74.922	N
Selenium	34	Se	78.96	N
Bromine	35	Br	79.904	N
Krypton	36	Kr	83.80	G
Rubidium	37	Rb	85.468	
Strontium	38	Sr	87.62	
Yttrium	39	Y	88.91	
Zirconium	40	Zr	91.224	
Niobium	41	Nb	92.906	
Molybdenum	42	Mo	95.94	
Technetium	43	Tc	[98]	
Ruthenium	44	Ru	101.1	
Rhodium	45	Rh	102.91	
Palladium	46	Pd	106.42	
Silver	47	Ag	107.87	
Cadmium	48	Cd	112.4	
Indium	49	In	114.8	
Tin	50	Sn	118.71	
Antimony	51	Sb	121.76	
Tellurium	52	Te	127.60	
Iodine	53	I	126.90	N
Xenon	54	Xe	131.29	G
Cesium	55	Cs	132.91	
Barium	56	Ba	137.33	
Lutetium	71	Lu	175	
Hafnium	72	Hf	178.49	
Tantalum	73	Ta	180.95	
Tungsten	74	W	183.84	
Rhenium	75	Re	186.2	
Osmium	76	Os	190.2	
Iridium	77	Ir	192.22	
Platinum	78	Pt	195.08	
Gold	79	Au	196.97	
Mercury	80	Hg	200.3	
Thallium	81	Tl	204.4	
Lead	82	Pb	207.2	
Bismuth	83	Bi	208.98	
Polonium	84	Po	[209]	
Astatine	85	At	[210]	
Radon	86	Rn	[222]	G
Francium	87	Fr	[223]	
Radium	88	Ra	[226]	
Lawrencium	103	Lr	[262]	
Rutherfordium	104	Rf	[262]	
Dubnium	105	Db	[262]	
Seaborgium	106	Sg	[266]	
Bohrium	107	Bh	[264]	
Hassium	108	Hs	[277]	
Meitnerium	109	Mt	[268]	
Darmstadtium	110	Ds	[281]	
Roentgenium	111	Rg	[272]	
Ununnilium	112	Uub	[285]	
Ununtrium	113	Uut	[284]	
Ununquadium	114	Uuq	[289]	
Ununpentium	115	Uup	[288]	
Ununhexium	116	Uuh	[292]	
Ununseptium	117	Uus	[?]	
Ununoctium	118	Uuo	[?]	

*Lanthanoids

Name	No.	Symbol	Atomic weight
Lanthanum	57	La	138.91
Cerium	58	Ce	140.12
Praseodymium	59	Pr	140.91
Neodymium	60	Nd	144.24
Promethium	61	Pm	[145]
Samarium	62	Sm	150.36
Europium	63	Eu	151.96
Gadolinium	64	Gd	157.25
Terbium	65	Tb	158.93
Dysprosium	66	Dy	162.50
Holmium	67	Ho	164.93
Erbium	68	Er	167.26
Thulium	69	Tm	168.93
Ytterbium	70	Yb	173.04

**Actinoids

Name	No.	Symbol	Atomic weight
Actinium	89	Ac	[227]
Thorium	90	Th	232.04
Protactinium	91	Pa	231.04
Uranium	92	U	238.03
Neptunium	93	Np	[237]
Plutonium	94	Pu	[244]
Americium	95	Am	[243]
Curium	96	Cm	[247]
Berkelium	97	Bk	[247]
Californium	98	Cf	[251]
Einsteinium	99	Es	[252]
Fermium	100	Fm	[257]
Mendelevium	101	Md	[258]
Nobelium	102	No	[259]

Printed and bound by CPI Group (UK) Ltd, Croydon, CR0 4YY

27/10/2024

14580296-0002